Density Matrix
Theory and
Applications

PHYSICS OF ATOMS AND MOLECULES

Series Editors:

P. G. Burke, *The Queen's University of Belfast, Northern Ireland and Daresbury Laboratory, Science Research Council, Warrington, England*

H. Kleinpoppen, *Institute of Atomic Physics, University of Stirling, Scotland*

1976:
ELECTRON AND PHOTON INTERACTIONS WITH ATOMS
Edited by H. Kleinpoppen and M.R.C. McDowell

1978:
PROGRESS IN ATOMIC SPECTROSCOPY, Parts A and B
Edited by W. Hanle and H. Kleinpoppen

1979:
ATOM–MOLECULE COLLISION THEORY: A Guide for the Experimentalist
Edited by Richard B. Bernstein

1980:
COHERENCE AND CORRELATION IN ATOMIC COLLISIONS
Edited by H. Kleinpoppen and J. F. Williams

VARIATIONAL METHODS IN ELECTRON–ATOM SCATTERING THEORY
R. K. Nesbet

1981:
DENSITY MATRIX THEORY AND APPLICATIONS
Karl Blum

INNER-SHELL AND X-RAY PHYSICS OF ATOMS AND SOLIDS
Edited by Derek J. Fabian, Hans Kleinpoppen, and Lewis M. Watson

---

A Continuation Order Plan is available for this series. A continuation order will bring delivery of each new volume immediately upon publication. Volumes are billed only upon actual shipment. For further information please contact the publisher.

Density Matrix Theory and Applications

Karl Blum

University of Munster
Federal Republic of Germany

Plenum Press • New York and London

Library of Congress Cataloging in Publication Data

Blum, Karl, 1937-
    Density matrix theory and applications.

    (Physics of atoms and molecules)
    Includes index.
    1. Density matrices. I. Title.
    QC174.17.D44B55   530.1'22   81-268
    ISBN 0-306-40684-5   AACR1

© 1981 Plenum Press, New York
A Division of Plenum Publishing Corporation
233 Spring Street, New York, N.Y. 10013

Printed in the United States of America

Preface

Quantum mechanics has been mostly concerned with those states of systems that are represented by state vectors. In many cases, however, the system of interest is incompletely determined; for example, it may have no more than a certain probability of being in the precisely defined dynamical state characterized by a state vector. Because of this incomplete knowledge, a need for statistical averaging arises in the same sense as in classical physics.

The density matrix was introduced by J. von Neumann in 1927 to describe statistical concepts in quantum mechanics. The main virtue of the density matrix is its analytical power in the construction of general formulas and in the proof of general theorems. The evaluation of averages and probabilities of the physical quantities characterizing a given system is extremely cumbersome without the use of density matrix techniques. The representation of quantum mechanical states by density matrices enables the maximum information available on the system to be expressed in a compact manner and hence avoids the introduction of unnecessary variables. The use of density matrix methods also has the advantage of providing a uniform treatment of all quantum mechanical states, whether they are completely or incompletely known.

Until recently the use of the density matrix method has been mainly restricted to statistical physics. In recent years, however, the application of the density matrix has been gaining more and more importance in many other fields of physics. For example, in modern atomic physics, density matrix techniques have become an important tool for describing the various quantum mechanical interference phenomena which are of importance in scattering theory, quantum beat spectroscopy, optical pumping, and laser physics. This book proposes to introduce the reader to the methods of density matrix theory with an emphasis on their application in atomic (and nuclear) physics. It is aimed at beginners and not experts. All the basic concepts are therefore discussed in detail and all the steps in the calculations are explained. As background, a standard one-year course in quantum mechanics is assumed, as is knowledge of the elements of statistical mechanics. Some background in modern atomic physics and scattering theory would also be helpful. For Chapters 4–6 the reader should have a working knowledge of angular momentum theory. Otherwise the treatment is begun from the lowest level possible. Some topics of contemporary interest are covered in sufficient detail to make the book also useful to those readers engaged in research in the fields of atomic or nuclear physics, laser physics, and physical chemistry.

The book can be divided into three main parts. In the first three chapters the basic concepts and methods of density matrix theory are introduced. In order to do this, some of the fundamental ideas of quantum mechanics and statistics are discussed. In particular, a clear understanding of pure and mixed quantum mechanical states is important. This is best achieved by considering simple systems. For this reason Chapter 1 is restricted to a discussion of the polarization states of spin-1/2 particles and photons, which enables all the basic concepts to be introduced in a simple way. The density matrix is first introduced as the counterpart of the distribution function of classical statistical mechanics, that is, by considering how many systems are in an ensemble with given wave functions. Then, after some of its basic properties are discussed, another aspect of the density matrix is considered: By introducing a convenient parametrization it is shown that the density matrix is the most convenient way of collecting all parameters of interest for a given experimental setup and of describing their behavior from an operational point of view.

In Chapter 2 these results are generalized to systems with more than two degrees of freedom and the basic properties of the density matrix derived in a more systematic way. The concept of coherence, which will be of central importance for the discussion of quantum mechanical interference phenomena in the following chapters is introduced. The properties of the time evolution operators are then reviewed and the basic equations of motion for statistical mixtures derived and illustrated with some examples.

In Chapter 3 another important aspect of the density matrix is introduced. Often, one is only interested in a few of many degrees of freedom of a quantum system, for example, when only one of several interacting systems is observed. In Sections 3.1 and 3.2 it is shown that, in general, it is impossible to find a wave function which depends only on the variables of the system of interest and not on those of all other systems as well. By averaging over all unobserved degrees of freedom a density matrix is obtained which describes the behavior of the system of interest. It is then shown that this "reduced" density matrix is the most general description of an open quantum mechanical system. The consequences of these general results are illustrated in Sections 3.3 and 3.4 with particular emphasis on the quantum mechanical theory of coherence. Finally, in Section 3.5 the reduced density matrix of atoms excited by electron impact is constructed and discussed in detail.

The subjects discussed in this chapter are related to the quantum mechanical theory of measurement. The questions raised here have attracted a great deal of interest from physicists in recent years.

The second part of the book (Chapters 4–6) is devoted to the application of the irreducible tensor method in density matrix theory. Quantum mechanical calculations for systems having symmetry can be divided into two parts. The first part consists of deriving as much information as possible from the symmetry requirements. The second part consists of calculating the dynamical quantities for which no information can be obtained from symmetry considerations. Often these two parts are tangled. The irreducible tensor method is designed to separate dynamical and geometrical elements and to provide a well-developed and efficient way of making use of the symmetry.

In Sections 4.2 and 4.3 the basic properties of tensor operators are discussed and the irreducible components of the density matrix (state multipoles, statistical tensors) are introduced. Sections 4.4–4.6 give various applications of the method, while Section 4.7 is devoted to a discussion of the time evolution of state multipoles in the presence of external perturbations.

The formalism developed up to this point is then applied in Chapters 5 and 6 to various problems of relevance to modern atomic spectroscopy, including the theory of quantum beats, electron–photon angular correlations, and the depolarization of the emitted radiation caused by fine and hyperfine interaction and magnetic fields. Throughout these chapters the discussion of quantum interference phenomena in atomic physics is based on the concept of "perturbation coefficients" developed by nuclear physicists in order to describe perturbed angular correlations. This formalism allows a very economic interpretation of experiments.

The last part of the book (Chapter 7) can be read independently of Chapters 4–6 (except some parts of Sections 7.5 and 7.6). In this chapter we discuss the density matrix approach to irreversible processes relating reversible and irreversible dynamics via generalized Master equations. Throughout this chapter the Markoff approximation is used. In Section 7.1 the fundamental concepts are introduced and the basic equations derived by considering the interaction between a "small" dynamic system and a "large" one ("heat bath"). Irreversibility is introduced by assuming that the bath remains in thermal equilibrium at constant temperature, irrespective of the amount of energy and information diffusing into it from the dynamic system. The special case of rate equations (Pauli's Master equation) is considered in

Section 7.2. The formalism is then applied to some simple examples in radio- and microwave spectroscopy. In order to illustrate the use of Master equation techniques in quantum electronics we consider the interaction between electromagnetic fields and two-level atoms. The corresponding Master equation is discussed in detail and the effects of relaxation inter- actions on the emitted line are described. In Section 7.4, the Bloch equations are derived and applied to magnetic resonance phenomena. It is shown that the density matrix method enables both longitudinal and transverse relax- ation to be treated in a natural way, thereby avoiding the shortcomings of semiclassical theories. The usefulness of the Bloch equations for a descrip- tion of the interaction between atoms or molecules and laser or maser fields is briefly considered.

The discussion of the general formalism is then continued by deriving the general properties of the relaxation matrix in Section 7.5. The discussion of the Liouville formalism in Section 7.6 is restricted to the basic concepts. Finally, in Section 7.7, the response of a quantum system to an external field is considered. Here, an expression is derived by approximating the exact equation of motion of the density matrix in retaining only terms linear in the field strength. This method is closely related to the theory of retarded Green's functions and is of importance for the investigation of transport phenomena.

The theory and application of the density matrix have been well summarized by various authors. Here, we mention in particular the review papers by Fano (1957) and ter Haar (1961). Some textbooks on quantum mechanics outline the formalism (Messiah, 1965; Roman, 1965; Gottfried, 1966). These sources (and many others which are acknowledged in the appropriate places) were used in writing this book. Because of the intro- ductory nature of this book we refer as a rule to monographs and reviews of the subject and only to those original papers whose results are used in the text.

Over the years my understanding of the theory and applications of the density matrix has benefited from many discussions with my colleagues at the Universities of Stirling and Münster. I am especially grateful to Prof. H. Kleinpoppen, who first aroused my interest in atomic physics, for his constant encouragement. I am indebted to Prof. J. Kessler for reading parts of the manuscript and making helpful suggestions for revisions in the first and second drafts. Dr. H. Jakubowicz has read the complete manuscript and made many improvements, and K. Bartschat has checked most equations. Finally, I wish to thank Mrs. Queen and Mrs. Raffin for their help in preparing the manuscript.

Karl Blum

Contents

Basic Concepts

1.1. Spin States and Density Matrix of Spin-1/2 Particles

1.1.1. Pure Spin States

In order to become familiar with the basic concepts of density matrix theory we will begin by considering the problem of describing the spin states of spin-1/2 particles. First of all we will review some results of the quantum mechanical theory of experiments with Stern–Gerlach magnets and then in the following sections we will reinterpret the results and discuss them in more detail.

Consider a beam of spin-1/2 particles (for example, hydrogen atoms) which passes through a Stern–Gerlach magnet which has its field gradient aligned along the z direction with respect to a fixed coordinate system x, y, z (Figure 1.1). In general the beam will split vertically into two parts each of which corresponds to one of the two possible eigenvalues of the component S_z of the spin operator $\mathbf{S}$ ($m = \pm 1/2$). If one of the beams is eliminated, for

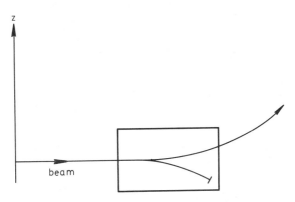

Figure 1.1. Stern–Gerlach filter.

example, the lower one as in Figure 1.1, then the emerging particles will be in a state which corresponds to only one of the eigenvalues; in the case of the apparatus in Figure 1.1 this would be $m = +1/2$. Similarly, if the apparatus is rotated in such a way that its field gradient points in the direction z', the emerging particles will be in a state which is described by the quantum number $m' = +1/2$, where m' is the eigenvalue of the operator $S_{z'}$, the component of $\mathbf{S}$ in the z' direction.

If the incident beam is such that it contains particles which are in a state with $m = +1/2$ only, then the beam will pass through the apparatus shown in Figure 1.1 without any loss of intensity. In all other cases part of the beam will be blocked off and the emerging beam will be less intense than the incident one. However, by tilting the apparatus at various angles about z it may be possible to find an orientation of the magnet which allows the whole beam to be transmitted. For example, if an incident beam contains only a spin component corresponding to $m' = +1/2$ in the frame z', it would be attenuated by the Stern–Gerlach apparatus in Figure 1.1. If the magnet were rotated so that its field gradient lay along z', then the beam would be completely transmitted. In this case all particles are deflected in the same way; they behave identically in this particular experiment. This enables the following (preliminary) definition to be made:

▶ If it is possible to find an orientation of the Stern–Gerlach apparatus for which a given beam is *completely* transmitted, then we will say that the beam is in a *pure spin state.*

In terms of the semiclassical vector model a beam of particles with definite quantum number $m = +1/2$ can be described by considering the spin vector of each particle to precess around the direction of the z axis such that its projection on the z axis has the value of $+1/2$ (Figure 1.2a). In this case the particles are said to have "spin up." A similar interpretation holds for the case of $m = -1/2$ (Figure 1.2b) and the spins of particles in eigenstates of the operator $S_{z'}$ will, by analogy, precess around the z' direction. In the case of a pure spin state the spin vectors of the particles precess around a unique direction which is parallel to the direction of alignment of the Stern–Gerlach apparatus when it allows the beam to pass through unattenuated.

If the state of a given beam is known to be pure then the joint state of all particles can be represented in terms of one and the same state vector $|\chi\rangle$. This is an important point and we will illustrate it with some examples. If a beam of particles passes completely through a Stern–Gerlach apparatus oriented in the z direction then we will say that *all* particles in the beam are in identical spin states with quantum number $m = 1/2$ with respect to z, or

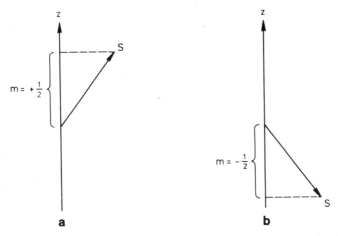

Figure 1.2. (a) Spin "in the z direction"; (b) spin "in the $-z$ direction."

that all particles have spin up with respect to z. We describe this state by assigning the state vector $|\chi\rangle = |+1/2\rangle$ to the *whole* beam. Similarly, a beam of particles with $m = -1/2$ will be characterized by $|\chi\rangle = |-1/2\rangle$. In the usual Pauli representation the state vectors are represented by two-dimensional column vectors,

$$\left|+\frac{1}{2}\right\rangle = \begin{pmatrix} 1 \\ 0 \end{pmatrix}; \qquad \left|-\frac{1}{2}\right\rangle = \begin{pmatrix} 0 \\ 1 \end{pmatrix} \tag{1.1.1a}$$

and the adjoint states by the row vectors

$$\langle +\tfrac{1}{2}| = (1, 0); \qquad \langle -\tfrac{1}{2}| = (0, 1) \tag{1.1.1b}$$

In general, for a beam that emerges from a Stern–Gerlach apparatus which has its magnet pointing in the z' direction, all particles in the beam are in a state with definite spin quantum number $m' = 1/2$ defined with respect to z' as quantization axis. The joint state of all particles will be described by the state vector $|\chi\rangle = |+1/2, z'\rangle$.

A general spin state $|\chi\rangle$ can always be written as a linear superposition of two basis states, for example, the states $|\pm 1/2\rangle$:

$$|\chi\rangle = a_1|+\tfrac{1}{2}\rangle + a_2|-\tfrac{1}{2}\rangle \tag{1.1.2}$$

In the representation (1.1.1) this is equivalent to

$$|\chi\rangle = \begin{pmatrix} a_1 \\ a_2 \end{pmatrix} \tag{1.1.2a}$$

The adjoint state is represented by the row vector

$$\langle\chi| = (a_1^*, a_2^*) \tag{1.1.2b}$$

where the asterisk denotes the complex conjugate.

The state $|\chi\rangle$ is normalized such that

$$\langle\chi|\chi\rangle = |a_1|^2 + |a_2|^2 = 1 \tag{1.1.3}$$

A pure spin state can be characterized either by specifying the direction in which the spins are pointing (for example, by giving the polar angles of this direction in our fixed coordinate system) or, alternatively, by specifying the coefficients a_1 and a_2 in the expansion (1.1.2). In the following section we will discuss the relation between these two descriptions and derive an explicit representation for the coefficients a_1 and a_2.

An apparatus of the type shown by Figure 1.1 acts as a *filter*, because irrespective of the state of the beam sent through it, the emerging beam is in a definite spin state which is defined by the orientation of the magnet. Passing a beam through the filter can therefore be regarded as a method of preparing a beam of particles in a pure state.

1.1.2. The Polarization Vector

In order to discuss the description of pure spin states in greater detail we will now introduce a vector **P**, called the *polarization vector*, which has components defined as expectation values of the corresponding Pauli matrices:

$$P_i = \langle\sigma_i\rangle \tag{1.1.4}$$

($i = x, y, z$). In the case of a pure state these expectation values are defined by the relations

$$\langle\sigma_i\rangle = \langle\chi|\sigma_i|\chi\rangle \tag{1.1.5}$$

In the representation (1.1.1) the Pauli matrices have the form

$$\sigma_x = \begin{pmatrix} 0 & 1 \\ 1 & 0 \end{pmatrix}, \qquad \sigma_y = \begin{pmatrix} 0 & -i \\ i & 0 \end{pmatrix}, \qquad \sigma_z = \begin{pmatrix} 1 & 0 \\ 0 & -1 \end{pmatrix} \tag{1.1.6}$$

The expectation values (1.1.5) may then be calculated by applying Eqs. (1.1.2a), (1.1.2b), and (1.1.6), treating the row and column vectors as one-dimensional matrices and applying the rules of matrix multiplication. In order to see the significance of the polarization vector we will now consider a few examples.

A beam of particles in the pure state $|+1/2\rangle$ has a polarization vector with components

$$P_x = (1, 0)\begin{pmatrix} 0 & 1 \\ 1 & 0 \end{pmatrix}\begin{pmatrix} 1 \\ 0 \end{pmatrix} = 0$$

$$P_y = (1, 0)\begin{pmatrix} 0 & -i \\ i & 0 \end{pmatrix}\begin{pmatrix} 1 \\ 0 \end{pmatrix} = 0 \qquad (1.1.7a)$$

$$P_z = (1, 0)\begin{pmatrix} 1 & 0 \\ 0 & -1 \end{pmatrix}\begin{pmatrix} 1 \\ 0 \end{pmatrix} = 1$$

Similarly we find for an ensemble of particles in the pure state $|-1/2\rangle$ the polarization vector has components

$$P_x = 0, \qquad P_y = 0, \qquad P_z = -1 \qquad (1.1.7b)$$

Thus the states $|+1/2\rangle$ and $|-1/2\rangle$ are characterized by polarization vectors of unit magnitude pointing in the $(+z)$- and $(-z)$-directions, respectively. The states $|+1/2\rangle$ and $|-1/2\rangle$ can therefore be said to be states of opposite polarization.

Consider now the general pure state (1.1.2). It will be convenient to give first of all a parametrization of the coefficients a_1 and a_2. These are complex numbers corresponding to four real parameters specifying the magnitudes and phases. The overall phase of the state (1.1.2) has no physical significance and can be chosen arbitrarily, for example by requiring a_1 to be real. From this condition and the normalization (1.1.3) it follows that the general pure spin state (1.1.2) is completely specified by two real numbers. It will therefore be convenient to introduce two parameters θ and δ defined by

$$a_1 = \cos\frac{\theta}{2}, \qquad a_2 = e^{i\delta}\sin\frac{\theta}{2} \qquad (1.1.8)$$

where δ is the relative phase of the coefficients. Using (1.1.8) Eq. (1.1.2a) becomes

$$|\chi\rangle = \begin{pmatrix} \cos\dfrac{\theta}{2} \\ e^{i\delta}\sin\dfrac{\theta}{2} \end{pmatrix} \qquad (1.1.9)$$

In order to see the physical significance of the parameters θ and δ consider the polarization vector associated with the state (1.1.9). We obtain

$$P_x = \left(\cos\frac{\theta}{2}, e^{-i\delta}\sin\frac{\theta}{2}\right)\begin{pmatrix}0 & 1\\ 1 & 0\end{pmatrix}\begin{pmatrix}\cos\dfrac{\theta}{2}\\ e^{i\delta}\sin\dfrac{\theta}{2}\end{pmatrix}$$

$$= \sin\theta\cos\delta \qquad\qquad (1.1.10a)$$

$$P_y = \sin\theta\sin\delta \qquad\qquad (1.1.10b)$$

$$P_z = \cos\theta \qquad\qquad (1.1.10c)$$

The polarization vector (1.1.10) has unit magnitude

$$|\mathbf{P}| = (P_x^2 + P_y^2 + P_z^2)^{1/2} = 1 \qquad\qquad (1.1.11)$$

From Eqs. (1.1.10) it follows that the parameters θ and δ can be interpreted as the polar angles of $\mathbf{P}$: θ is the angle between $\mathbf{P}$ and the z axis and the relative phase δ specifies the azimuth angle of $\mathbf{P}$ (Figure 1.3).

A second coordinate system x', y', z' can be chosen in such a way that the z' axis is parallel to $\mathbf{P}$. Taking z' as quantization axis we have in the primed system $P_{x'} = 0$, $P_{y'} = 0$, $P_{z'} = 1$, that is, all particles have spin up with respect to z'. The direction of the polarization vector is therefore the direction "in which the spins are pointing." If we send the beam through a Stern–Gerlach filter oriented parallel to $\mathbf{P}$ the whole beam will pass through.

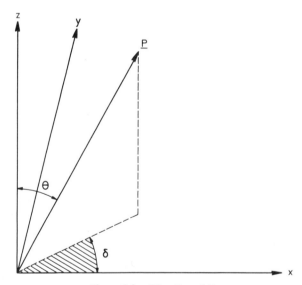

Figure 1.3. Direction of $\mathbf{P}$.

Equations (1.1.9) and (1.1.10) enable explicit spin functions to be constructed for any pure state. For example, a given beam of particles may be in a pure state with spins pointing in the x direction of our fixed coordinate system. In this case the corresponding polarization vector points in the x direction and, consequently, has polar angles $\theta = 90°$, $\delta = 0$. From Eq. (1.1.9) the state vector can be seen to be

$$|+\tfrac{1}{2}, x\rangle = \frac{1}{2^{1/2}}\binom{1}{1} \qquad (1.1.12a)$$

A beam of particles with "spin down" with respect to the x axis has a polarization vector pointing in the $(-x)$ direction and is characterized by the angles $\theta = 90°$, $\delta = 180°$. The state vector is represented by

$$|-\tfrac{1}{2}, x\rangle = \frac{1}{2^{1/2}}\binom{1}{-1} \qquad (1.1.12b)$$

Similarly, the state vector of particles with "spin up" ("spin down") with respect to the y axis is represented by the column vectors

$$|+\tfrac{1}{2}, y\rangle = \frac{1}{2^{1/2}}\binom{1}{i} \qquad (1.1.12c)$$

and

$$|-\tfrac{1}{2}, y\rangle = \frac{1}{2^{1/2}}\binom{1}{-i} \qquad (1.1.12d)$$

It should be noted that the four states (1.1.12) are constructed by superposing the states $|+1/2\rangle$ and $|-1/2\rangle$ with the same magnitude $|a_1| = |a_2| = 1/2^{1/2}$ but with different relative phases. The corresponding polarization vectors have the same angle θ but different azimuth angles depending on the relative phase existing between the states $|\pm 1/2\rangle$.

1.1.3. Mixed Spin States

Pure spin states are not the most general spin states in which an ensemble of particles can be found. Suppose, for example, that two beams of particles have been prepared *independently*, one in the pure state $|+1/2\rangle$ and the other one in the pure state $|-1/2\rangle$. By "independently" we mean that no definite phase relation exists between the two beams (this point will be clarified later). The first beam may consist of N_1 particles, the second one of N_2 particles. If the polarization state of the combined beam is investigated by sending it through a Stern–Gerlach filter in various orientations it will be found that it is not possible to find such an orientation of the filter which allows the whole beam to be transmitted. It follows that, by definition, the joint beam is not in a pure spin state.

► States which are not pure are called *mixed states* or *mixtures*.

We now have to consider the problem of describing the state of the joint beam. Clearly, it is not possible to characterize the state of the beam in terms of a single state vector $|\chi\rangle$ since associated with any of these states there is necessarily a direction in which *all* spins are pointing: the direction of the polarization vector. If the Stern–Gerlach filter were placed in this orientation the whole beam would have to be transmitted. Since no such orientation exists it is not possible to describe a mixture by a single state vector.

In particular, the mixture cannot be represented by a linear super-position of the states $|+1/2\rangle$ and $|-1/2\rangle$ representing the two constituent beams. In order to construct such a linear superposition it is necessary to know the magnitudes and relative phase δ of the relevant coefficients a_1 and a_2. The absolute squares $|a_1|^2$ and $|a_2|^2$ are probabilities W_1 and W_2 of finding a particle in the state $|+1/2\rangle$ and $|-1/2\rangle$, respectively. In the case of the mixture under discussion these probabilities are known ($W_1 = N_1/N$ and $W_2 = N_2/N$ with $N = N_1 + N_2$) and can be used to determine the magnitudes of the coefficients ($W_1 = |a_1|^2$, $W_2 = |a_2|^2$). The important point is that the constituent beams have been prepared independently. So there is no definite phase relation between the two beams and without a definite phase δ it is not possible to construct a well-defined state vector $|\chi\rangle$ describing the joint beam.

A mixture has to be described by specifying the way in which it has been prepared. For example, the joint beam under discussion is characterized by saying that N_1 particles have been prepared in the state $|+1/2\rangle$ and N_2 in the state $|-1/2\rangle$ independently of each other. This statement contains all the information we have obtained about the mixture.

Let us continue discussion of our example by calculating the polariza-tion vector associated with the total beam. **P** is obtained by taking the statistical average over the separate beams:

$$P_i = W_1 \langle \tfrac{1}{2} | \sigma_i | \tfrac{1}{2} \rangle + W_2 \langle -\tfrac{1}{2} | \sigma_i | -\tfrac{1}{2} \rangle$$

which gives

$$P_x = 0, \qquad P_y = 0, \qquad P_z = W_1 - W_2 = \frac{N_1 - N_2}{N} \qquad (1.1.13)$$

It should be noted that the polarization vector (1.1.13) has a magnitude which is less than 1 and is proportional to the difference of the population numbers of the two states, $|+1/2\rangle$ and $|-1/2\rangle$.

More generally, if a beam is prepared by mixing N_a particles in the state $|\chi_a\rangle$ and N_b in the state $|\chi_b\rangle$ then the components of the polarization vector are given by the statistical average over the independently prepared beams:

$$P_i = W_a\langle\chi_a|\sigma_i|\chi_a\rangle + W_b\langle\chi_b|\sigma_i|\chi_b\rangle \qquad (1.1.14)$$

$$= W_a P_i^{(a)} + W_b P_i^{(b)} \qquad (1.1.14a)$$

with $W_a = N_a/N$, $W_b = N_b/N$, and where $P_i^{(a)}$ and $P_i^{(b)}$ are the polarization vectors associated with the constituent beams [see, Eq. (1.1.5)]. Equation (1.1.14) can be rewritten in vector notation as

$$\mathbf{P} = W_a\mathbf{P}^{(a)} + W_b\mathbf{P}^{(b)} \qquad (1.1.14b)$$

Since $|P^{(a)}| = 1$, $|P^{(b)}| = 1$ the magnitude of P is determined by

$$
\begin{aligned}
P^2 &= (W_a\mathbf{P}^{(a)} + W_b\mathbf{P}^{(b)})^2 \\
&= W_a^2(P^{(a)})^2 + W_b(P^{(b)})^2 + 2W_aW_b\mathbf{P}^{(a)}\cdot\mathbf{P}^{(b)} \\
&\leq W_a^2 + W_b^2 + 2W_aW_b \\
&= (W_a + W_b)^2 = 1 \qquad (1.1.15)
\end{aligned}
$$

Since the scalar product $\mathbf{P}^{(a)}\cdot\mathbf{P}^{(b)}$ of two different unit vectors is less than 1.

The equality sign in Eq. (1.1.15) applies if $\mathbf{P}^{(a)}\cdot\mathbf{P}^{(b)} = 1$, that is, if the two beams have identical polarization vectors. In this case both constituent beams are in the same spin state described by Eqs. (1.1.9) and (1.1.10) and the joint beam is in a pure state. Vice versa, if two beams are mixed in identical spin states the resulting beam consists of particles in identical spin states and is therefore characterized by a polarization vector of unit magnitude. The above arguments may be easily generalized to cases of mixtures which consist of more than two beams.

We therefore have the following result: The magnitude of the polarization vector is bounded such that

$$0 \leq |P| \leq 1 \qquad (1.1.16)$$

The maximum possible value $|P| = 1$ is obtained if (and only if) the beam under consideration is in a pure state, whereas mixtures necessarily have a polarization vector of less than unit magnitude.

This result once more illustrates the basic property of a pure spin state: all the particles are in identical states with *all* the spins pointing in the same direction, the direction of **P**.

Henceforward we will refer to states with $|P| > 0$ as "*polarized*" and to beams with $|P| = 0$ as "*unpolarized.*" Pure states ($|P| = 1$) will be called "*completely polarized.*"

1.1.4. Pure versus Mixed States

Before proceeding further with any analysis it is important to have a clear understanding of the concepts of pure and mixed states. We will therefore consider both types of states again from a different viewpoint. Consider the following problem. A beam of particles which is completely polarized in the y direction and hence can be represented by the state vector (1.1.12c)

$$|+\tfrac{1}{2}, y\rangle = \frac{1}{2^{1/2}}(|+\tfrac{1}{2}\rangle + i|-\tfrac{1}{2}\rangle) \qquad (1.1.17a)$$

is sent through a Stern–Gerlach filter oriented in the z direction. What will happen? It is a familiar result of quantum mechanics that, although we know that any particle in the beam is in the state $|+1/2, y\rangle$, it is impossible to predict whether a given single particle will pass through the filter. This is because a system which is to be measured is in general disturbed by the act of measurement. In this case the measuring apparatus (the filter) changes in a completely uncontrollable way the state of the incident particles, that is, it is only possible to predict the probability that a particle will be admitted by the filter (and emerge in the state $|+1/2\rangle$) or be blocked off. From Eq. (1.1.17a) it can be seen that the probability is $1/2$ for each case. The only case in which it is possible to predict with complete certainty whether a given particle will pass through a filter or not is that where the filter is oriented in the y direction since then all particles will pass unhindered. In general, however, the measuring process can only be described through the use of statistics.

Because of this a linear superposition state, such as (1.1.17a), must be interpreted as follows. Before any measurement all the particles are in identical states represented by the vector (1.1.17a) and all particles have the same quantum number $m' = 1/2$ defined with respect to y as the quantization axis. The quantum number m, defined with respect to the z axis, is completely undefined in the superposition (1.1.17a) in the sense that *any* particle in the beam has an equal probability of passing through a z-oriented filter or being blocked off. [Roughly speaking, it can be said that the particles in the superposition state (1.1.17a) do not "know" their m value.] If the beam is sent through a filter oriented parallel to the z axis the interaction with the apparatus changes the state of the beam and forces the particles into one of the eigenstates.

Consider now a mixture of

$$N_1 = N/2 \text{ particles in the state } |+1/2\rangle$$
$$N_2 = N/2 \text{ particles in the state } |-1/2\rangle \qquad (1.1.17b)$$

with both subbeams prepared independently. From Eq. (1.1.13) it can be seen that the resulting beam is unpolarized. If this beam is passed through a Stern–Gerlach filter oriented along the z axis the transmitted beam will have half of the incident intensity. In this particular experiment the mixture (1.1.17b) and the pure state (1.1.17a) give the same result; however, it is for a different reason. Whereas in the case of the state $|1/2, y\rangle$ all particles in the beam are in one and the same state, there is less information about the mixture (1.1.17b) since it is only known that any particle has an equal probability of being in the state $|+1/2\rangle$ or $|-1/2\rangle$. In this sense the state of the mixture is *incompletely determined*. When passing through the filter the particles with $m = -1/2$ will be blocked off and hence only that half of the beam corresponding to the $|+1/2\rangle$-component beam will be transmitted.

The above example illustrates that statistics must be used in order to describe the *initial* state of the mixture; the state of the particles is not known with certainty, that is, we cannot assign a single-state vector to the mixed beam.

In conclusion, it can be seen that in the description of spin-1/2 particles statistics enters in two ways. First of all, statistical methods must be used because of the uncontrollable perturbation of states by any measuring apparatus. Secondly, when dealing with mixtures, it is only known that the particles can be in any one of several spin states. A statistical description must be applied because of the lack of information available on the system. It was primarily for the purpose of describing this latter case that the density matrix formalism was developed.

A more systematic treatment of the problems discussed above will be presented in Chapter 2.

1.1.5. The Spin-Density Matrix and Its Basic Properties

1.1.5.1. Basic Definitions

Any question concerning the behavior of pure or mixed states can be answered by specifying the states, present in the mixture, and their statistical weights W_i. The actual calculations, however, are often very cumbersome. We will therefore now introduce an alternative method of characterizing pure and mixed states.

Consider a beam of N_a particles prepared in the state $|\chi_a\rangle$ and a second beam of N_b particles which have been prepared in the state $|\chi_b\rangle$ independently of the first one. In order to describe the joint beam we introduce an operator ρ by the expression

$$\rho = W_a|\chi_a\rangle\langle\chi_a| + W_b|\chi_b\rangle\langle\chi_b| \tag{1.1.18}$$

with $W_a = N_a/N$, $W_b = N_b/N$, and $N = N_a + N_b$.

The operator ρ is called *density operator* or *statistical operator*. It describes the preparations which have been performed and, therefore, contains all the information obtained on the total beam. In this sense a mixture is completely specified by its density operator. In the special case of a pure state $|\chi\rangle$ the density operator is given by

$$\rho = |\chi\rangle\langle\chi| \tag{1.1.18a}$$

It will be seen later that it is usually more convenient to write ρ in matrix form. To this end we choose a set of basis states, commonly $|+1/2\rangle$ and $|-1/2\rangle$ and expand $|\chi_a\rangle$ and $|\chi_b\rangle$ in terms of this set according to Eq. (1.1.2):

$$\begin{aligned}|\chi_a\rangle &= a_1^{(a)}|+\tfrac{1}{2}\rangle + a_2^{(a)}|-\tfrac{1}{2}\rangle \\ |\chi_b\rangle &= a_1^{(b)}|+\tfrac{1}{2}\rangle + a_2^{(b)}|-\tfrac{1}{2}\rangle\end{aligned} \tag{1.1.19}$$

In the representation (1.1.1) we write

$$|\chi_a\rangle = \begin{pmatrix} a_1^{(a)} \\ a_2^{(a)} \end{pmatrix}, \qquad |\chi_b\rangle = \begin{pmatrix} a_1^{(b)} \\ a_2^{(b)} \end{pmatrix} \tag{1.1.20a}$$

and for the adjoint states

$$\langle\chi_a| = \left(a_1^{(a)*}, a_2^{(a)*}\right), \qquad \langle\chi_b| = (a_1^{(b)*}, a_2^{(b)*}) \tag{1.1.20b}$$

Applying the rules of matrix multiplication we obtain for the "outer product," $|\chi_a\rangle\langle\chi_a|$:

$$\begin{aligned}|\chi_a\rangle\langle\chi_a| &= \begin{pmatrix} a_1^{(a)} \\ a_2^{(a)} \end{pmatrix}(a_1^{(a)*}, a_2^{(a)*}) \\ &= \begin{pmatrix} |a_1^{(a)}|^2 & a_1^{(a)}a_2^{(a)*} \\ a_1^{(a)*}a_2^{(a)} & |a_2^{(a)}|^2 \end{pmatrix}\end{aligned} \tag{1.1.21}$$

and similarly for the product $|\chi_b\rangle\langle\chi_b|$. Substitution of these expressions into Eq. (1.1.18) yields the *density matrix*

$$\rho = \begin{pmatrix} W_a|a_1^{(a)}|^2 + W_b|a_1^{(b)}|^2 & W_a a_1^{(a)}a_2^{(a)*} + W_b a_1^{(b)}a_2^{(b)*} \\ W_a a_1^{(a)*}a_2^{(a)} + W_b a_1^{(b)}a_2^{(b)*} & W_a|a_2^{(a)}|^2 + W_b|a_2^{(b)}|^2 \end{pmatrix} \tag{1.1.22}$$

Since the basis states $|\pm 1/2\rangle$ have been used in deriving Eq. (1.1.22) this is said to be *the density matrix in the $\{|\pm 1/2\rangle\}$ representation*.

In order to make subsequent formulas more compact we define $|+1/2\rangle = |\chi_1\rangle$ and $|-1/2\rangle = |\chi_2\rangle$. In this notation the general element of the density matrix corresponding to the ith row and jth column is given by the expression

$$\langle \chi_i | \rho | \chi_j \rangle = W_a a_i^{(a)} a_j^{(a)*} + W_b a_i^{(b)} a_j^{(b)*} \tag{1.1.23}$$

with $i, j = 1, 2$.

Clearly the density matrix has a different form in different representations, whereas the operator (1.1.18) is independent of the choice of the basis states. It will always be assumed that the basis states are orthonormal, that is

$$\langle \chi_i | \chi_j \rangle = \delta_{ij} \tag{1.1.24}$$

where δ_{ij} denotes the Kronecker symbol and for $i = j$ condition (1.1.3) is satisfied.

In the normalization (1.1.3) the trace of the density matrix is given by

$$\mathrm{tr}\, \rho = W_a + W_b = 1 \tag{1.1.25}$$

which is independent of the choice of the representation.

As an example, consider the case of a mixture consisting of N_1 particles which have been prepared in the state $|\chi_1\rangle = |+1/2\rangle$ and N_2 particles prepared independently in the state $|\chi_2\rangle = |-1/2\rangle$. The total beam is then represented by the density operator

$$\rho = W_1 |+\tfrac{1}{2}\rangle\langle +\tfrac{1}{2}| + W_2 |-\tfrac{1}{2}\rangle\langle -\tfrac{1}{2}| \tag{1.1.26a}$$

($W_i = N_i/N$) and the density matrix in the $\{|\pm 1/2\rangle\}$ representation is diagonal:

$$\langle \chi_i | \rho | \chi_j \rangle = W_i \delta_{ij} \tag{1.1.26b}$$

1.1.5.2. Significance of the Density Matrix

The diagonal elements of the density matrix

$$\langle \chi_i | \rho | \chi_i \rangle = W_a |a_i^{(a)}|^2 + W_b |a_i^{(b)}|^2 \qquad (i = 1, 2) \tag{1.1.27}$$

have a direct physical meaning. Since the probability of finding a particle of the mixture in the state $|\chi_a\rangle$ is W_a and since the probability that $|\chi_a\rangle$ is in the state $|\chi_i\rangle$ is $|a_i^{(a)}|^2$, the product $W_a |a_i^{(a)}|^2$ is the probability that a particle originally prepared in the state $|\chi_a\rangle$ will be found in the state $|\chi_i\rangle$ after a measurement has been made. *The diagonal element (1.1.27) therefore gives the total probability of finding a particle in the corresponding basis state $|\chi_i\rangle$.*

Thus, if a beam described by a density operator ρ is sent through a Stern–Gerlach filter oriented parallel (antiparallel) to the z axis then the diagonal element $\langle \chi_1 | \rho | \chi_1 \rangle = \langle +\frac{1}{2} | \rho | +\frac{1}{2} \rangle$ $(\langle -\frac{1}{2} | \rho | -\frac{1}{2} \rangle)$ of ρ in the $\{|\pm\frac{1}{2}\rangle\}$ representation gives the probability that a particle will pass through the filter.

This result may be generalized to arbitrary states $|\chi\rangle$. Consider the matrix element $\langle \chi | \rho | \chi \rangle$ obtained by "sandwiching" the operator (1.1.18) between the state $|\chi\rangle$ and its adjoint $\langle \chi |$:

$$\langle \chi | \rho | \chi \rangle = W_a \langle \chi | \chi_a \rangle \langle \chi_a | \chi \rangle + W_b \langle \chi | \chi_b \rangle \langle \chi_b | \chi \rangle$$
$$= W_a |a^{(a)}|^2 + W_b |a^{(b)}|^2 \qquad (1.1.28)$$

where $a^{(a)} = \langle \chi_a | \chi \rangle$ and $a^{(b)} = \langle \chi_b | \chi \rangle$. Comparing Eqs. (1.1.27) and (1.1.28) it can be seen that the *matrix element* $\langle \chi | \rho | \chi \rangle$ *is the total probability of finding a particle in the pure state* $|\chi\rangle$ *within a mixture which is represented by* ρ. That is, if a beam represented by ρ passes through a filter which only fully admits a beam in the state $|\chi\rangle$ then Eq. (1.1.28) gives the probability that any given particle of the beam will pass through the filter.

For example, suppose that a beam represented by the density matrix (1.1.26) is sent through a filter oriented in y direction. The probability that a particle of the beam will pass through is then given by the matrix element $\langle +1/2, y | \rho | +1/2, y \rangle$. Expressing $|+1/2, y\rangle$ in the $\{|\pm 1/2\rangle\}$ representation [Eq. (1.1.12c)] and using Eq. (1.1.26) gives

$$\langle +\tfrac{1}{2}, y | \rho | +\tfrac{1}{2}, y \rangle = \tfrac{1}{2}(1, -i) \begin{pmatrix} W_1 & 0 \\ 0 & W_2 \end{pmatrix} \begin{pmatrix} 1 \\ i \end{pmatrix}$$
$$= \tfrac{1}{2}(W_1 + W_2)$$

The important point to be noted is that all the information on the spin state of any given beam can be obtained (in principle at least) by sending the beam through various Stern–Gerlach filters with different orientations. Consequently, once ρ is known, we can calculate the result of any such experiment by means of Eq. (1.1.28). In this sense, ρ *contains all significant information on the spin state of a given beam*.

1.1.5.3. The Number of Independent Parameters

We will now consider how many parameters are required in order to completely represent a given density matrix. A complex 2×2 matrix such as (1.1.22) has four complex elements $\langle \chi_i | \rho | \chi_j \rangle$ corresponding to eight real parameters. The density matrix in *Hermitian*, that is, ρ satisfies

$$\langle \chi_i | \rho | \chi_j \rangle = \langle \chi_j | \rho | \chi_i \rangle^* \qquad (1.1.29)$$

This can be seen immediately from Eq. (1.1.22) or (1.1.23). Consequently, the diagonal elements are real and furthermore the real and imaginary parts of the off-diagonal elements are related by the expressions

$$\text{Re} \langle +\tfrac{1}{2}|\rho|-\tfrac{1}{2}\rangle = \text{Re} \langle -\tfrac{1}{2}|\rho|+\tfrac{1}{2}\rangle$$

$$\text{Im} \langle +\tfrac{1}{2}|\rho|-\tfrac{1}{2}\rangle = -\text{Im} \langle -\tfrac{1}{2}|\rho|+\tfrac{1}{2}\rangle$$

These relations reduce the number of independent real parameters to four. The normalization condition (1.1.25) fixes one further parameter so that the density matrix is completely characterized in terms of three real parameters. It follows from this that *three independent measurements must be performed in order to completely specify the density matrix for any given beam of spin-1/2 particles.*

It will be instructive to consider this result from another point of view. In Eq. (1.1.18) a density operator is defined from a knowledge of how a given beam has been prepared. This definition can be generalized for the case of any number of constituent beams. In order to write down the density operator or the corresponding density matrix from Eqs. (1.1.18) and (1.1.22), it follows that all the pure states $|\chi_a\rangle, |\chi_b\rangle, \cdots$ present in the mixture must be specified together with their statistical weights $W_a, W_b, \cdots$. Only three real parameters, however, are required to completely characterize the density matrix of a beam of any complexity, as has been shown above.

This is not as surprising as it may seem at first, since one and the same density matrix can represent many different mixtures prepared in entirely different ways. For example, consider a mixture represented by the density operator

$$\rho = \tfrac{1}{2}|+\tfrac{1}{2}\rangle\langle+\tfrac{1}{2}| + \tfrac{1}{2}|-\tfrac{1}{2}\rangle\langle-\tfrac{1}{2}|$$

and a mixture specified by the operator

$$\rho = \tfrac{1}{6}|+\tfrac{1}{2}\rangle\langle+\tfrac{1}{2}| + \tfrac{1}{6}|-\tfrac{1}{2}\rangle\langle-\tfrac{1}{2}| + \tfrac{1}{3}|+\tfrac{1}{2}, x\rangle\langle+\tfrac{1}{2}, x| + \tfrac{1}{3}|-\tfrac{1}{2}, x\rangle\langle-\tfrac{1}{2}, x|$$

By constructing the corresponding density matrices in the $\{\langle\pm1/2|\}$ representation and applying Eqs. (1.1.12a) and (1.1.12b) it can be shown that both beams are represented by the same density matrix:

$$\rho = \frac{1}{2}\begin{pmatrix} 1 & 0 \\ 0 & 1 \end{pmatrix}$$

It follows from Eq. (1.1.28) that the two beams will behave identically in all experiments with respect to their polarization properties. Conversely, a knowledge of the density matrix elements alone is insufficient to determine the method by which the beams have been prepared. In fact such information is insignificant. The only significant information is contained in the three independent parameters specifying the density matrix since these are sufficient to calculate the behavior of the corresponding beam in any polarization experiment. *For this reason we will henceforth consider two beams to be identical if they are described by the same density matrix.*

The definition (1.1.18) is usually of little importance, and instead of defining the density *operator* by specifying the constituent subbeams and their statistical weights we will apply a more operational point of view and define the density *matrix* by the results of three independent measurements. In the following section we will show how this can be achieved in a simple way using the polarization vector.

1.1.5.4. Parametrization of the Density Matrix

If Eq. (1.1.18) is multiplied by the Pauli matrix σ_i the trace can be calculated as

$$
\begin{aligned}
\text{tr}\, \rho \sigma_i &= W_a\, \text{tr}\, (|\chi_a\rangle\langle\chi_a|\sigma_i) + W_b\, \text{tr}\, (|\chi_b\rangle\langle\chi_b|\sigma_i) \\
&= W_a\langle\chi_a|\sigma_i|\chi_a\rangle + W_b\langle\chi_b|\sigma_i|\chi_b\rangle
\end{aligned}
\tag{1.1.30}
$$

This result can be obtained by using the explicit matrix representations (1.1.6) and (1.1.21) or, more directly, by applying the relation

$$
\text{tr}\, (|\chi\rangle\langle\chi|\sigma_i) = \langle\chi|\sigma_i|\chi\rangle
\tag{1.1.31}
$$

Substituting Eq. (1.1.14a) into Eq. (1.1.30) gives the important result

▶
$$
\text{tr}\, \rho \sigma_i = P_i
\tag{1.1.32}
$$

where P_i is the ith component of the polarization vector of the total beam.

Using this result the elements of ρ can be expressed in terms of the components P_i. By direct matrix manipulations it may be shown that in the $\{|\pm 1/2\rangle\}$ representation ρ is given by

▶
$$
\rho = \frac{1}{2}\begin{pmatrix} 1 + P_z & P_x - iP_y \\ P_x + iP_y & 1 - P_z \end{pmatrix}
\tag{1.1.33}
$$

A more elegant method of deriving this result will be given in Section 1.1.6.

The three components P_x, P_y, P_z represent a minimum set of data which are required to specify the density matrix of any given beam and *we will henceforth regard the density matrix as defined by Eq. (1.1.33).*

As an illustration of the use of Eq. (1.1.33) suppose a beam of particles, characterized by the matrix (1.1.33), passes a filter oriented in the z direction. The probability that a particle is admitted by the filter is given, according to Eq. (1.1.28), by the expression

$$\langle +\tfrac{1}{2}|\rho|+\tfrac{1}{2}\rangle = \tfrac{1}{2}(1 + P_z)$$

Similarly, applying Eqs. (1.1.12a), (1.1.12c), and (1.1.33) the probabilities that a particle will pass through a filter oriented in the x and y directions are found to be

$$\langle +\tfrac{1}{2}, x|\rho|+\tfrac{1}{2}, x\rangle = \tfrac{1}{2}(1 + P_x)$$

$$\langle +\tfrac{1}{2}, y|\rho|+\tfrac{1}{2}, y\rangle = \tfrac{1}{2}(1 + P_y)$$

Finally, we will give another useful representation of ρ obtained by transforming to a coordinate system x', y', z', where z' is parallel to $\mathbf{P}$ and x' and y' are chosen arbitrarily but orthogonal to each other and to z'. In this case, $P_{x'} = P_{y'} = 0, P_{z'} = |P|$. As a result, in the representation with z' as the quantization axis, ρ is given by

$$\rho = \frac{1}{2}\begin{pmatrix} 1 + |\mathbf{P}| & 0 \\ 0 & 1 + |P| \end{pmatrix} \tag{1.1.34a}$$

or alternatively, by

$$\rho = \tfrac{1}{2}(1 - |P|)\begin{pmatrix} 1 & 0 \\ 0 & 1 \end{pmatrix} + |P|\begin{pmatrix} 1 & 0 \\ 0 & 0 \end{pmatrix} \tag{1.1.34b}$$

If the beam under consideration is completely polarized, $|P| = 1$, and

$$\rho = \begin{pmatrix} 1 & 0 \\ 0 & 0 \end{pmatrix} \tag{1.1.35}$$

and the beam is in the pure state $|+1/2, z'\rangle$. If the beam is unpolarized $|P| = 0$, and the corresponding density matrix is given by

$$\rho = \frac{1}{2}\begin{pmatrix} 1 & 0 \\ 0 & 1 \end{pmatrix} \tag{1.1.36}$$

1.1.5.5. Identification of Pure States

In Section 1.1.2 it has been shown that a given beam is in a pure state if and only if its polarization vector has the maximum possible value of $|P| = 1$. This result will now be put into a different form which is more useful for treating more complex systems.

Using Eq. (1.1.33) it can be shown that the trace of ρ^2 is given by

$$\text{tr}\,(\rho^2) = (1/2)(1 + P_x^2 + P_y^2 + P_z^2)$$

$$= (1/2(1 + |P|^2)$$

It follows from this that

$$\text{tr}\,(\rho^2) = 1 \qquad (1.1.37)$$

is a necessary and sufficient condition that the beam under consideration is in a pure state. [Note, the fact that the trace is equal to unity in Eq. (1.1.37) is a consequence of the normalization (1.1.25).]

In the case of a pure state the condition (1.1.37) gives an additional restriction on the density matrix elements. Thus *a pure state is characterized by two independent parameters only in accordance with Eq. (1.1.9).*

1.1.6. The Algebra of the Pauli Matrices

The discussions in Section 1.1.5 have shown that the result of any experiment performed with a given beam can be calculated from a knowledge of the corresponding density matrix. So far the required mathematical operations have to be carried out using a particular representation and applying the rules of matrix algebra. In general, this is a laborious and time-consuming procedure. In this section a more elegant method of performing the relevant calculations will be described.

The discussion will be based on the following fundamental relation between the Pauli matrices $(i, j = x, y, z)$:

$$\sigma_i \sigma_j = \delta_{ij} \mathbf{1} + i \sum_k \varepsilon_{ijk} \sigma_k \qquad (1.1.38)$$

where δ_{ij} is the Kronecker symbol, $\mathbf{1}$ denotes the two-dimensional unit matrix, and

$$\varepsilon_{ijk} = \begin{cases} 1 & \text{if } i, j, k \text{ is an even permutation of } XYZ \\ -1 & \text{if } i, j, k \text{ is an odd permutation of } XYZ \\ 0 & \text{if two of the indices are the same} \end{cases} \qquad (1.1.39)$$

For example, for $i = j$ Eq. (1.1.38) becomes

$$\sigma_i^2 = 1 \qquad (1.1.40a)$$

and for $i = x, j = y$:

$$\sigma_x \sigma_y = i\sigma_z, \qquad \sigma_y \sigma_x = -i\sigma_z \qquad (1.1.40b)$$

From Eqs. (1.1.37) and (1.1.40) it follows that for $i \neq j$

$$\sigma_i \sigma_j + \sigma_j \sigma_i = 0 \qquad (1.1.40c)$$

Equation (1.1.38) specifies completely the algebra of the Pauli matrices. Proofs of Eq. (1.1.38) can be found in any of the textbooks on quantum mechanics.

The important property of Eq. (1.1.38) is that it reduces quadratic combinations of Pauli matrices to linear ones. This allows the calculation of traces of products of matrices σ_i by a stepwise reduction of the number of matrices occurring in the given trace. We give some examples. First, from Eq. (1.1.6) it can be seen that

$$\text{tr } \sigma_i = 0 \qquad (1.1.41)$$

By taking the trace of Eq. (1.1.38) and using Eq. (1.1.41) it follows that

$$\text{tr } \sigma_i\sigma_j = 2\delta_{ij} \qquad (1.1.42a)$$

A product of three Pauli matrices may first be reduced to a quadratic combination by means of Eq. (1.1.38):

$$\sigma_i\sigma_j\sigma_m = \delta_{ij}\sigma_m + i \sum_k \varepsilon_{ijk}\sigma_k\sigma_m$$

Taking the trace of this expression and applying Eqs. (1.1.41) and (1.1.42a) gives

$$\text{tr } \sigma_i\sigma_j\sigma_m = 2i \sum_k \varepsilon_{ijk}\delta_{km} = 2i\varepsilon_{ijm} \qquad (1.1.42b)$$

A further important property of the Pauli matrices is that any two-dimensional Hermitian matrix can be expressed as a linear combination of the unit matrix $\mathbf{1}$ and the matrices σ_i. For example, consider the density matrix. We make the "ansatz"

$$\rho = a\mathbf{1} + \sum_i b_i\sigma_i \qquad (1.1.43)$$

In Eq. (1.1.43) the four coefficients a, b_x, b_y, b_z are unknowns which must be determined. Such an ansatz is possible because the Hermiticity condition reduces the number of independent parameters determining ρ to four and there are four parameters in Eq. (1.1.41). One of the parameters can immediately be determined from the normalization condition (1.1.25) which gives with the help of Eq. (1.1.41)

$$a = 1/2 \qquad (1.1.44a)$$

Multiplying Eq. (1.1.43) by σ_j, taking the trace of the obtained expression, and using Eqs. (1.1.41) and (1.1.42) gives

$$\text{tr } \rho\sigma_j = 2 \sum_i b_i\delta_{ij}$$

$$= 2b_j$$

The trace of ρ and σ_j gives the corresponding component of the polarization vector and from this it follows that

$$b_j = (1/2)P_j \qquad (1.1.44b)$$

Inserting the results (1.1.44) into the ansatz (1.1.43) results in the expression

▶ $$\rho = \frac{1}{2}\left(1 + \sum_i P_i\sigma_i\right) \qquad (1.1.45)$$

If the Pauli matrices are expressed in the form (1.1.6) then P can be obtained in the form (1.1.33). In the case of a pure state $|\chi\rangle$ characterized by

$$\rho^{(\chi)} = |\chi\rangle\langle\chi|$$

then, denoting the polarization vector of the state $|\chi\rangle$ by $P^{(\chi)}$, we write

$$|\chi\rangle\langle\chi| = \frac{1}{2}\left(1 + \sum_i P_i^{(\chi)}\sigma_i\right) \qquad (1.1.46)$$

This expression allows a simple determination of the probability $\langle\chi|\rho|\chi\rangle$. Equation (1.1.31) implies

$$\langle\chi|\rho|\chi\rangle = \mathrm{tr}\,|\chi\rangle\langle\chi|\rho$$

Hence, using this result on the right-hand side of Eq. (1.1.46) gives

$$\langle\chi|\rho|\chi\rangle = \tfrac{1}{4}\mathrm{tr}\left[\left(1 + \sum_i (P_i^{(\chi)}\sigma_i)\right)\left(1 + \sum_j P_j\sigma_j\right)\right]$$

$$= \tfrac{1}{4}\mathrm{tr}\left(1 + \sum_i P_i^{(\chi)}\sigma_i + \sum_j P_j\sigma_j + \sum_{i,j} P_i^{(\chi)}P_j\sigma_i\sigma_j\right)$$

$$= \tfrac{1}{2}(1 + \mathbf{P}^{(\chi)} \cdot \mathbf{P}) \qquad (1.1.47)$$

This result can be interpreted in the following way. A beam of particles may be characterized by a density matrix ρ. This beam may be passed through a Stern–Gerlach filter in a fixed orientation which only completely transmits a beam in the pure state $|\chi\rangle$ (that is, the filter is oriented parallel to $\mathbf{P}^{(\chi)}$).

The probability that a particle of the given beam will pass through the filter is then determined by the scalar product $\mathbf{P}^{(\chi)} \cdot \mathbf{P}$ of the two polarization vectors. The probability of transmission is a maximum if $\mathbf{P}$ points in the direction of the magnetic field and is a minimum if $\mathbf{P}$ is antiparallel to the filter direction. In particular, if the beam is unpolarized, then for *any* filter

$$\langle\chi|\rho|\chi\rangle = 1/2 \qquad (1.1.48)$$

The derivation of Eq. (1.1.47) may serve as a first example of how calculations can be simplified by using Eq. (1.1.45) and the algebraic properties of the Pauli matrices.

1.1.7. Summary

The results obtained in the previous two sections allow a redefinition of the basic concepts used so far. We consider as the initial information on a given beam the values of the three components P_x, P_y, P_z of the polarization vector. **P** can be determined, for example, by suitably chosen scattering experiments (for a detailed discussion of such experiments we refer particularly to Kessler, 1976). When the polarization vector is known the density matrix can be obtained by means of Eqs. (1.1.33) or (1.1.45). These expressions contain all information on the beam in condensed form. The usefulness particularly of Eq. (1.1.45) in actual calculations will become evident in Section 2.5.

If $|P| = 1$ the beam is said to be in a pure spin state, or, alternatively, all particles are in identical states. This joint state of all particles in the given beam is represented by assigning a single-state vector to the whole beam. In this case two parameters are sufficient for a complete description of the spin state, for example, the polar angles θ and δ of **P**, from which the corresponding state vector can be constructed by means of Eq. (1.1.9).

If $|P| < 1$ the beam is said to be in a mixed state. Such states are characterized by three parameters, for example, the magnitude and the polar angles of **P**.

1.2. Polarization States and Density Matrix of Photons

1.2.1. The Classical Concept of Wave Polarization

In this section, a description of photon polarization will be given. We will follow the arguments of Section 1.1 in order to become more familiar with the abstract concepts introduced there. We will begin with a brief account of the description of light polarization in classical optics.

A monochromatic electromagnetic wave is characterized by three quantities: its angular frequency ω, its wave vector $\mathbf{k} = (2\pi/\Lambda)\mathbf{n}$ (where **n** is a unit vector in the direction of motion and Λ is the wavelength), and its state of polarization, which is defined by the vibrations of the electric field vector **E**. The field vector **E** of a monochromatic wave can be written in the form

$$\mathbf{E} = A\mathbf{e}\, e^{i(\mathbf{k}\cdot\mathbf{r}-\omega t)} \qquad (1.2.1)$$

where A is the amplitude and **e** the polarization vector. Because of the transverse nature of electromagnetic waves, **e** is perpendicular to **n**. In this section we will use a coordinate system x, y, z with the z axis parallel to **n**, and restrict ourselves to a discussion of the polarization properties of light only. If **E** vibrates along the x axis then the light is said to be linearly polarized along the x axis. The polarization vector is parallel to x and denoted by $\mathbf{e}_x$. If the electric vector oscillates along the y axis then the polarization is characterized by assigning a polarization vector $\mathbf{e}_y$ to the beam pointing in the y'direction. A general polarization vector $\mathbf{e}$ can always be expanded in terms of two orthogonal vectors, for example, $\mathbf{e}_x$ and $\mathbf{e}_y$:

$$\mathbf{e} = a_1 \, \mathbf{e}_x + a_2 \, e^{i\delta} \, \mathbf{e}_y \qquad (1.2.2)$$

where a_1 and a_2 are real coefficients. We will normalize Eq. (1.2.2) such that $\mathbf{e}$ is always a unit vector in the sense that the scalar product of $\mathbf{e}$ and its complex conjugate, $\mathbf{e}^*$, is equal to 1: $\mathbf{e} \cdot \mathbf{e}^* = 1$. The normalization condition is therefore

$$a_1^2 + a_2^2 = 1 \qquad (1.2.3)$$

Equation (1.2.2) corresponds to a linear superposition of two waves of equal frequency and the same wave vector with amplitudes A_1 and A_2, polarized along the x and y directions, respectively, with a definite phase difference δ:

$$\mathbf{E} = A_1 \mathbf{e}_x \, e^{i(\mathbf{k}\cdot\mathbf{r}-\omega t)} + A_2 \mathbf{e}_y \, e^{i(\mathbf{k}\cdot\mathbf{r}-\omega t+\delta)}$$

$$= A(a_1\mathbf{e}_x + a_2 e^{i\delta}\mathbf{e}_y) \, e^{i(\mathbf{k}\cdot\mathbf{r}-\omega t)}$$

where a_1 and a_2 are the relative amplitudes of the waves normalized to unity: $a_i = A_i/A$ $(i = 1, 2)$ with $A = (A_1 + A_2)^{1/2}$.

We can define a parameter β such that

$$a_1 = \cos\beta, \qquad a_2 = \sin\beta \qquad (1.2.4a)$$

[Eq. (1.2.3) is then automatically satisfied] and write the general polarization vector (1.2.2) in the form

▶ $$\mathbf{e} = \cos\beta \, \mathbf{e}_x + e^{i\delta} \sin\beta \, \mathbf{e}_y \qquad (1.2.4b)$$

In order to become familiar with the use of this expression we will consider some specific cases.

1. Consider a superposition of two waves oscillating in phase, with relative amplitudes a_1 and a_2 and polarized along the x and y axes, respectively. From the relative amplitudes the parameter β can be determined, and inserting $\delta = 0$ into Eq. (1.2.4) the polarization vector of the resulting wave is found to be

$$\mathbf{e} = \cos\beta \, \mathbf{e}_x + \sin\beta \, \mathbf{e}_y \qquad (1.2.5)$$

In this case it is possible to give a simple interpretation of β: $\mathbf{e}$ is a real vector in the x–y plane and Eq. (1.2.5) represents its decomposition in terms of the two orthogonal basis vectors $\mathbf{e}_x$ and $\mathbf{e}_y$; hence, β is the angle between e and the x axis (Figure 1.4).

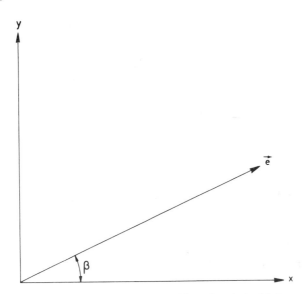

Figure 1.4. Polarization vector of linearly polarized light.

2. A superposition of two waves with equal amplitudes $a_1 = a_2$ and a phase difference $\delta = \pm 90°$ gives a wave with polarization vector

$$\mathbf{e} \sim \mathbf{e}_x \pm i\,\mathbf{e}_y$$

corresponding to left- and right-handed circular polarization. (For a further discussion of circular polarization see Section 1.2.3.)

3. If $a_1 \neq a_2$ and $\delta \neq 0$ we have the general case of elliptical polarization.

In the following we will refer to a light wave as *completely polarized if its polarization properties can be specified in terms of a single polarization vector* $\mathbf{e}$ [as in the case of the plane wave (1.1.1), for example]. It will be useful to reinterpret this definition in terms of some idealized experiments.

Following a treatment similar to that given in Section 1.1 the polarization properties of light can be discussed with the help of experiments with various optical polarization filters. We will assume that the filters used are always ideal in the sense that the filter is completely transparent to light of a particular polarization and completely absorbs light of the opposite polarization. Hence light passing through the filter will emerge in a definite state of polarization. For example, a beam of light may pass a Nicol prism with its axis of transmission parallel to the x axis. The transmitted light is then linearly polarized along the x direction. Similarly, a beam of light passing a Nicol prism oriented parallel to an axis $\mathbf{n}$ will emerge linearly polarized along this direction. If β is the angle between **n** and the x axis the corresponding polarization vector is given by Eq. (1.2.5). Conversely, if linearly polarized light with polarization vector **e** is passed through a Nicol prism it is always possible to find such an orientation of the prism which allows the whole beam to be transmitted. This occurs when the axis of transmission is parallel to **e**. A circularly polarized wave will only be completely accepted by a circular polarization filter (for example, a suitably oriented series combination of a quarter wave plate and a Nicol prism).

By applying the converse of these arguments it can be seen that a light beam is completely polarized if such a filter can be found which completely admits the beam.

As is well known from optics light is usually not completely polarized. An ordinary light source consists of a large number of excited atoms each of which emits a pulse of light in a time of order $\sim 10^{-8}$ sec independently of all other atoms. Because constantly new pulses will be contributing to the beam the overall polarization will change very rapidly and there will be no definite polarization vector which is characteristic of the total beam. In the following sections we will consider the problem of describing beams of this kind.

1.2.2. Pure and Mixed Polarization States of Photons

When the theory of relativistic quantum mechanics is applied to the electromagnetic field it follows that in interaction with matter the wave behaves as if it were composed of photons. We will start our discussion with the following definition:

► A beam of photons is said to be in a *pure polarization state* if the beam is completely polarized in the sense explained in Section 1.2.1.

In terms of our idealized experiments this definition may be reinterpreted as follows: if it is possible to find such a filter which completely admits a beam of photons then the beam is said to be in a pure polarization state. Alternatively we may say that all photons of the beam can be considered to be in one and the same polarization state. This joint state of *all* the photons can be described in terms of a *single* state vector which we will denote by $|e\rangle$, by which is meant the polarization state of any photon in the beam which classically has polarization vector **e**. For example, the state vectors $|e_x\rangle$ and $|e_y\rangle$ denote the polarization state of photons which are completely transmitted by a Nicol prism orientated in the x and y directions, respectively.

The states $|e_x\rangle$ and $|e_y\rangle$ can be taken as basis states and any state $|e\rangle$ can be written as a linear superposition:

$$|e\rangle = a_1|e_x\rangle + a_2|e_y\rangle \qquad (1.2.6)$$

or

$$|e\rangle = \cos\beta |e_x\rangle + e^{i\delta} \sin\beta |e_y\rangle \qquad (1.2.7)$$

These equations are exactly analogous to Eqs. (1.2.2) and (1.2.4), respectively.

These considerations are similar to the discussions in Section 1.1. All the experiments and results for spin-1/2 particles and Stern–Gerlach filters which have been described previously can be repeated with photons and polarization filters. In particular, it should be noted that a_1^2 and a_2^2 are the probabilities that a photon in the polarization state (1.2.6) will pass through a Nicol prism oriented parallel to the x or y axis, respectively.

As shown by Eqs. (1.2.6) and (1.2.7) any superposition of two (or more) states which have a definite phase δ necessarily results in a pure state. Thus, in order to describe light which is not completely polarized, it is necessary to consider superposition states which do not have a definite phase relation, that is, we have to introduce the concept of a mixture. In general, *a beam of photons is said to be in a mixed state or mixture if it is not possible to describe the beam in terms of a single state vector.*

It is useful to visualize the concept of a mixture in terms of some idealized experiments. Consider two light sources emitting *independently* of each other, where by "independently" is meant that there is no definite phase relationship between the two sources (that is, the relative phase changes much more rapidly than the observation time in an unpredictable manner). Both sources are provided with a polarization filter so that the first source emits a beam of intensity I_1 of definite polarization $|e_1\rangle$ and the second one a beam of intensity I_2 and polarization $|e_2\rangle$. If the two beams are combined and the polarization properties of the total beam investigated by sending it through various filters, it will be found that, irrespective of the nature of the filter, the transmitted intensity is always less than the incident one. Thus, by definition, the total beam is in a mixed polarization state.

It is not possible to completely characterize a mixture by a single-state vector $|e\rangle$. In particular, the mixture cannot be represented as a linear superposition of the states $|e_1\rangle$ and $|e_2\rangle$. The reason for this is that, as has been discussed in Sections 1.1.1–1.1.4, there is no definite phase δ between the constituent beams with which a definite state vector $|e\rangle$ can be constructed.

1.2.3. The Quantum Mechanical Concept of Photon Spin

In classical optics the polarization of light is explained in terms of the vibrations of the electric field vector. We will now investigate how polarization states can be interpreted in terms of the characteristic properties of photons.

To this end we will consider the possible spin states of photons. There are certain limitations to the concept of the photon spin. The total angular momentum $\mathbf{J}$ of any particle is the resultant of its spin $\mathbf{S}$ and its orbital angular momentum $\mathbf{L}$. Since the rest mass of a photon is zero, the usual definition of spin as the total angular momentum of a particle at rest is inapplicable for a photon. Strictly, only the total angular momentum $\mathbf{J}$ of the photon has any physical meaning. However, it is convenient to define a spin and an orbital angular momentum in a formal sense. The photon spin is given the value 1 corresponding to the fact that the wave function is a vector [as shown, for example, by Eq. (1.2.1)]. The value of the orbital angular momentum is related to the multipoles which occur in the wave function (see, for example, Landau and Lifschitz, 1965).

In general, if a spin-1 particle has a well-defined momentum $\mathbf{p}$ the components of its spin along its direction of motion can take three values: $+1, -1, 0$. However, because of the transverse nature of electromagnetic waves the value 0 must be excluded for photons. The component of the photon spin along the direction of propagation $\mathbf{n}$, which we will denote by the symbol λ, can therefore only have the values $\lambda = +1$ ("spin up") and $\lambda = -1$ ("spin down").

It is important to note that the two photon states with spin up and spin down with respect to $\mathbf{n}$ as quantization axis have direct physical meaning. Since the component of the orbital angular momentum vanishes in the direction of propagation $\mathbf{n}$ we have $\mathbf{J} \cdot \mathbf{n} = (\mathbf{L} + \mathbf{S}) \cdot \mathbf{n} = \mathbf{S} \cdot \mathbf{n} = \lambda$; consequently,

▶ λ is the component of the total angular momentum of the photon in the direction of propagation $\mathbf{n}$.

The component of the spin in the direction of motion is generally called the helicity and we will refer to photon states with definite values $\lambda = \pm 1$ as *helicity states*.

Classically, when a circularly polarized light beam is directed at a target the electrons in the target are set into circular motion in response to the rotating electric field of the wave. This suggests that there is a relationship between circularly polarized light and photons in definite states of angular momentum.

In fact it has been shown in quantum electrodynamics *that photons with definite helicity are related to left-handed and right-handed circular polariza-tion states*. Unfortunately, this notation is not unambiguous and we will adopt the following convention. We will denote the polarization vector and state of photons with helicity $\lambda = 1$ by $\mathbf{e}_{+1}$ and $|+1\rangle$, respectively, and to refer to light of positive helicity as *right-handed circularly polarized*.

Similarly, if $\lambda = -1$ we will denote polarization vector and state of such light by $\mathbf{e}_{-1}$ and $|-1\rangle$, respectively, and refer to the light as *left-handed circularly polarized.*

Note, that in the terminology of classical optics the opposite convention is usually adopted: Light of positive (negative) helicity is called left-handed (right-handed) circularly polarized. We will always use the helicity state notion in order to avoid this ambiguity. With this convention the vectors $\mathbf{e}_{+1}$ and $\mathbf{e}_{-1}$ and the state states $|\pm 1\rangle$ are then determined apart from a phase factor which has little significance and we write

$$\mathbf{e}_{\pm 1} = \mp \frac{1}{2^{1/2}}(\mathbf{e}_x \pm i\,\mathbf{e}_y) \tag{1.2.8}$$

for the polarization vector and for the corresponding states:

$$|\pm 1\rangle = \mp \frac{1}{2^{1/2}}(|e_x\rangle \pm i|e_y\rangle) \tag{1.2.9}$$

(see, for example, Messiah, 1965).

In particular for problems where questions of angular momentum must be taken explicitly into account, it is convenient to use the helicity states as basis states instead of $|e_x\rangle$ and $|e_y\rangle$. We will therefore write the general polarization state $|e\rangle$ in the form

$$|e\rangle = a_1|+1\rangle + a_2|-1\rangle \tag{1.2.10}$$

There is a close formal analogy between photons and spin-1/2 particles. Because there are only two possible values of the helicity $\lambda = \pm 1$ (corresponding to states with spin up and spin down with respect to $\mathbf{n}$ as the quantization axis) these states can be represented by two-dimensional column vectors *as long as* $\mathbf{n}$ *is used as the axis of quantization* (z axis). The basis states can then be written in a similar way to Eqs. (1.1.1):

$$|+1\rangle = \begin{pmatrix} 1 \\ 0 \end{pmatrix}, \qquad |-1\rangle = \begin{pmatrix} 0 \\ 1 \end{pmatrix} \tag{1.2.11}$$

In this representation the general pure state (1.2.9) is described by the row vector

$$|e\rangle = \begin{pmatrix} a_1 \\ a_2 \end{pmatrix} \tag{1.2.11a}$$

and its adjoint by the column vector

$$\langle e| = (a_1^*, a_2^*) \tag{1.2.11b}$$

For example, the state of light beams which are completely linearly polarized along the x and y axes, respectively, is obtained by inverting Eq. (1.2.9):

$$|e_x\rangle = -\frac{1}{2^{1/2}}(|+1\rangle - |-1\rangle) \tag{1.2.12a}$$

$$|e_y\rangle = \frac{i}{2^{1/2}}(|+1\rangle + |-1\rangle) \tag{1.2.12b}$$

The interpretation of these linear superposition states is analogous to that given in Section 1.1.4.

As another example, consider a beam of photons prepared in the pure state $|e_x\rangle$. It can be seen from Eq. (1.2.12a) that these photons have no definite helicity. However, in any experiment performed on the beam, in which the angular momentum is actually measured, any photon in the beam will be forced into one of the angular momentum eigenstates, $|+1\rangle$ or $|-1\rangle$, with equal probability. In any such experiment any photon of the beam will therefore transfer a definite amount of angular momentum, either $\lambda = +1$ or $\lambda = -1$. Since the corresponding probabilities are equal, the net angular momentum, transferred by the total beam, is zero.

1.2.4. The Polarization Density Matrix

A compact expression of the polarization properties of photons is contained in the corresponding density matrix. In Section 1.2.5 we will give an operational definition of the photon density matrix. But in this section we will follow the arguments of Section 1.1.5.

Consider a beam of photons which is a mixture of two beams which have been prepared independently in the states $|e_a\rangle$ and $|e_b\rangle$, with intensities I_a and I_b, respectively. The density operator characterizing the total beam is defined by the expression

$$\rho' = W_a|e_a\rangle\langle e_a| + W_b|e_b\rangle\langle e_b| \tag{1.2.13}$$

with $W_a = I_a/I$ and $W_b = I_b/I$ and $I = I_a + I_b$. In order to obtain the density matrix it is necessary to choose a particular representation. We will use the helicity states as basis and expand the two states $|e_a\rangle$ and $|e_b\rangle$ according to Eq. (1.2.10) as

$$|e_a\rangle = a_1^{(a)}|+1\rangle + a_2^{(a)}|-1\rangle$$

$$|e_b\rangle = a_1^{(b)}|+1\rangle + a_2^{(b)}|-1\rangle$$

Using the explicit representation (1.2.11) and applying the rule (1.1.21) the density matrix in the helicity representation is found to be

$$\rho' = \begin{pmatrix} W_a|a_1^{(a)}|^2 + W_b|a_1^{(b)}|^2 & W_a a_1^{(a)} a_2^{(a)*} + W_b a_1^{(b)} a_2^{(b)*} \\ W_a a_1^{(a)*} a_2^{(a)} + W_b a_1^{(b)*} a_2^{(b)} & W_a|a_2^{(a)}|^2 + W_b|a_2^{(b)}|^2 \end{pmatrix} \tag{1.2.14}$$

From the explicit representation (1.2.14) it follows that ρ' is normalized according to

$$\text{tr}\,\rho' = W_a + W_b = 1 \tag{1.2.15}$$

It is often more convenient to normalize ρ in such a way that its trace is equal to the total intensity of the corresponding photon beam. This can be achieved by substituting the intensities I_a and I_b for W_a and W_b in the definition (1.2.13) and Eq. (1.2.14). The density operator in this normalization is then given by

$$\rho = I_a |e_a\rangle\langle e_a| + I_b |e_b\rangle\langle e_b| \tag{1.2.16}$$

The trace of the density matrix is then

$$\text{tr}\,\rho = I_a + I_b = I \tag{1.2.17}$$

The density matrix ρ (and ρ') has the following properties (the proofs are like those in Section 1.1.5):

1. The diagonal elements $\langle+1|\rho'|+1\rangle$ and $\langle-1|\rho'|-1\rangle$ of the matrix (1.2.14) give the probability of finding a photon in the beam in the corresponding helicity state. In the normalization (1.2.17) the diagonal elements $\langle+1|\rho|+1\rangle$ and $\langle-1|\rho|-1\rangle$ give the corresponding intensities.

2. If the beam under consideration is sent through a filter which fully admits only photons in the pure state $|e\rangle$ then the element

$$\langle e|\rho'|e\rangle = W_a |a^{(a)}|^2 + W_b |a^{(b)}|^2 \tag{1.2.18}$$

gives the probability that a photon of the beam will be transmitted through the filter where we used the notation

$$a^{(a)} = \langle e|e_a\rangle$$
$$a^{(b)} = \langle e|e_b\rangle$$

The element $\langle e|\rho|e\rangle$ obtained from the operator (1.2.16) gives the transmitted intensity:

$$\langle e|\rho|e\rangle = I_a |a^{(a)}|^2 + I_b |a^{(b)}|^2 \tag{1.2.19}$$

Since any information on the polarization properties of a given beam can be obtained in principle by allowing the beam to pass through various polarization filters, the result of any such experiment can be calculated by using formulas (1.2.18) or (1.2.19). From this it can be concluded that all information on the polarization state of a given beam is contained in its density matrix.

3. The hermiticity condition (1.1.29) reduces the number of independent parameters to four. One of these is usually the total beam intensity I. If I is not of interest it can be dropped by normalizing as in Eq. (1.2.14), where ρ'

then is specified by three real parameters similar to the density matrix of spin-1/2 particles.

It follows that four independent measurements are required in order to specify the matrix ρ completely for any given beam (one of these is the determination of the intensity I). If I is dropped the matrix (1.2.14), which requires a set of three independent measurements, is obtained. The result of any further experiment can then be calculated by applying Eqs. (1.2.18) or (1.2.19).

4. In Chapter 2 we will prove that, in general, a necessary and sufficient condition, that a given photon density matrix describes a pure state, is given by

$$\text{tr} \,(\rho^2) = (\text{tr} \,\rho)^2 = I^2 \qquad (1.2.20)$$

In the normalization (1.2.15) this reduces to Eq. (1.1.37):

$$\text{tr} \,(\rho^2) = 1$$

In general, the photon density matrix satisfies

$$\text{tr} \,(\rho^2) \le I^2 \qquad (1.2.20a)$$

1.2.5. Stokes Parameter Description

1.2.5.1. Parametrization of ρ in Terms of the Stokes Parameters

We will henceforth adopt the normalization (1.2.17). In the preceding section it has been seen that four independent measurements must be performed in order to completely determine the polarization state of any given beam. The most convenient set of measurements is that one which gives the following information:

1. The total intensity I of the beam;
2. the degree of linear polarization with respect to the x and y axes, defined as

$$\eta_3 = \frac{I(0) - I(90°)}{I} \qquad (1.2.21a)$$

where $I(\beta)$ denotes the intensity transmitted by a Nicol prism oriented at an angle β with respect to the x axis;

3. the degree of linear polarization with respect to two orthogonal axes oriented at 45° to the x axes

$$\eta_1 = \frac{I(45°) - I(135°)}{I} \qquad (1.2.21b)$$

4. the degree of circular polarization defined as

$$\eta_2 = \frac{I_{+1} - I_{-1}}{I} \tag{1.2.21c}$$

where I_{+1} (and I_{-1}) are the intensities of light transmitted by polarization filters which fully transmit only photons with positive (negative) helicity.

The set of the parameters 1–4 above is called *Stokes parameters*. A detailed description of these can be found in Born and Wolf (1970). (See also McMaster, 1954, and Farago, 1971.)

We will now relate the Stokes parameters to the elements of the density matrix. Denoting the elements of ρ by $\rho_{\lambda'\lambda} \equiv \langle \lambda' | \rho | \lambda \rangle$ we write

$$\rho = \begin{pmatrix} \rho_{+1,+1} & \rho_{+1,-1} \\ \rho_{-1,+1} & \rho_{-1,-1} \end{pmatrix} \tag{1.2.22}$$

where $\rho_{+1,-1} = \rho^*_{-1,+1}$ because of the Hermiticity condition (1.1.29). Applying Eq. (1.2.17) the total intensity is given by

$$I = \rho_{11} + \rho_{-1,-1} \tag{1.2.23a}$$

In order to obtain η_3 we have to calculate the intensities $I(0)$ and $I(90°)$. From relation (1.2.19) these are given by

$$I(0) = \langle e_x | \rho | e_x \rangle$$

$$I(90°) = \langle e_y | \rho | e_y \rangle$$

In the helicity representation the state vectors $|e_x\rangle$ and $|e_y\rangle$ are expressed by Eq. (1.2.12), hence

$$I(0) = (1/2)(-1, +1)\begin{pmatrix} \rho_{11} & \rho_{1,-1} \\ \rho_{-1,1} & \rho_{-1,-1} \end{pmatrix}\begin{pmatrix} -1 \\ +1 \end{pmatrix}$$

$$= (1/2)(\rho_{11} - \rho_{1,-1} - \rho_{-1,1} + \rho_{-1,-1})$$

Similarly, we obtain

$$I(90°) = (1/2)(-i, -i)\begin{pmatrix} \rho_{11} & \rho_{1,-1} \\ \rho_{-1,1} & \rho_{-1,-1} \end{pmatrix}\begin{pmatrix} i \\ i \end{pmatrix}$$

$$= (1/2)(\rho_{11} + \rho_{1,-1} + \rho_{-1,1} + \rho_{-1,-1})$$

It therefore follows that

$$I\eta_3 = -(\rho_{1,-1} + \rho_{-1,1}) \tag{1.2.23b}$$

In the same way we calculate the parameter $I\eta_1$, defined by Eq. (1.2.21c). In this case the axes of transmission of the Nicols are set at angles $45°$ and $135°$ to the x axes, respectively. The intensities transmitted by these prisms are then given by

$$I(45°) = \langle e_1|\rho|e_1\rangle$$
$$I(135°) = \langle e_2|\rho|e_2\rangle$$

where $|e_1\rangle$ denotes a photon state which is fully transmitted by the first prism, that is,

$$|e_1\rangle = (1/2^{1/2})(|e_x\rangle + |e_y\rangle)$$

where Eq. (1.2.5) has been used with $\beta = 45°$. Similarly, $|e_2\rangle$ is a photon state which is fully transmitted by the second prism and can be expressed in terms of $|e_x\rangle$ and $|e_y\rangle$ by inserting $\beta = 135°$ in Eq. (1.2.5):

$$|e_2\rangle = (1/2^{1/2})(-|e_x\rangle + |+|e_y\rangle)$$

Transforming $|e_x\rangle$ and $|e_y\rangle$ to the helicity basis gives

$$I\eta_1 = -i(\rho_{1,-1} - \rho_{-1,1}) \tag{1.2.23c}$$

Similarly,

$$I\eta_2 = \rho_{11} - \rho_{-1,-1} \tag{1.2.23d}$$

By inverting these equations the elements $\rho_{\lambda'\lambda}$ can be expressed in terms of the Stokes parameters:

▶
$$\rho = \frac{I}{2}\begin{pmatrix} 1 + \eta_2 & -\eta_3 + i\eta_1 \\ -\eta_3 - i\eta_1 & 1 - \eta_2 \end{pmatrix} \tag{1.2.24}$$

We will use this form of the density matrix throughout this book.

1.2.5.2. Examples

It follows from Eq. (1.2.7) that any pure polarization state can be parametrized in the form

$$|e\rangle = \cos\beta|e_x\rangle + e^{i\delta}\sin\beta|e_y\rangle \tag{1.2.25}$$

The corresponding density operator is given by $\rho = I|e\rangle\langle e|$. We will calculate the Stokes parameters characterizing a beam in the state (1.2.25). We have

$$I(0) = \langle e_x|\rho|e_x\rangle = I|\langle e_x|e\rangle|^2 = I\cos^2\beta$$
$$I(90°) = \langle e_y|\rho|e_y\rangle = I|\langle e_y|e\rangle|^2 = I\sin^2\beta$$

from which follows

$$\eta_3 = \cos 2\beta \tag{1.2.26a}$$

Similarly it is found

$$\eta_1 = \sin 2\beta \cos \delta \tag{1.2.26b}$$

$$\eta_2 = \sin 2\beta \sin \delta \tag{1.2.26c}$$

For example, the pure state $|e_x\rangle$, characterizing a beam of light with polarization vector pointing in the x direction, is obtained by inserting $\delta = 0$, $\beta = 0$ in Eq. (1.2.25). From Eqs. (1.2.26) we obtain the Stokes parameters $\eta_3 = 1$, $\eta_1 = \eta_2 = 0$. Inserting these values into the density matrix (1.2.24) gives

$$\rho = \frac{I}{2}\begin{pmatrix} 1 & -1 \\ -1 & 1 \end{pmatrix} \tag{1.2.27a}$$

A beam which is linearly polarized in the y direction can be specified by the parameters $\beta = 90°$, $\delta = 0$ so that

$$\eta_3 = -1, \qquad \eta_1 = \eta_2 = 0$$

and

$$\rho = \frac{I}{2}\begin{pmatrix} 1 & i \\ -i & 1 \end{pmatrix} \tag{1.2.27b}$$

Similarly, as shown in Section 1.2.1 a beam linearly polarized in a direction with angle θ with respect to the x axis is described by inserting $\delta = 0$ in Eqs. (1.2.25) and (1.2.26). The Stokes parameters are therefore given by $\eta_3 = \cos 2\beta$, $\eta_1 = \sin 2\beta$, $\eta_2 = 0$, and the corresponding density matrix is

$$\rho = \frac{I}{2}\begin{pmatrix} 1 & -\cos 2\beta + i \sin 2\beta \\ -\cos 2\beta - i \sin 2\beta & 1 \end{pmatrix} \tag{1.2.27c}$$

Left- and right-handed polarized light is represented by the density matrices

$$\rho = I\begin{pmatrix} 1 & 0 \\ 0 & 0 \end{pmatrix} \tag{1.2.28a}$$

and

$$\rho = I\begin{pmatrix} 0 & 0 \\ 0 & 1 \end{pmatrix} \tag{1.2.28b}$$

respectively.

Once the Stokes parameters, and hence the density matrix, have been determined, it is straightforward to derive a useful expression for the intensity, I_e, of light transmitted by a filter which only admits photons in the state $|e\rangle$: The required element is $\langle e|\rho|e\rangle$, which is

▶ $$I_e = (I/2)(1 + \eta_3 \cos 2\beta + \eta_1 \sin 2\beta \cos \delta + \eta_2 \sin 2\beta \sin \delta) \tag{1.2.29}$$

Note that the parameters β and δ describe the transmitted beam whereas the incident beam is specified in terms of the Stokes parameters in Eq. (1.2.29).

1.2.5.3. Degree of Polarization

We will now introduce a further notation which will be useful in later discussions. From condition (1.2.20a) and Eq. (1.2.24) it follows that the Stokes parameters are restricted by the condition

$$\eta_1^2 + \eta_2^2 + \eta_3^2 \leq 1 \qquad (1.2.30)$$

The equality sign holds only if the photons in the beam under discussion are in a pure polarization state. Alternatively, the beam is completely polarized (in the sense explained in Section 1.2.1) if and only if the relation

$$\eta_1^2 + \eta_2^2 + \eta_3^2 = 1 \qquad (1.2.30a)$$

holds. These conditions may be conveniently expressed by introducing the quantity

$$P = (\eta_1^2 + \eta_2^2 + \eta_3^2)^{1/2} \qquad (1.2.31)$$

It follows from Eq. (1.2.46a) that P is restricted by

$$P \leq 1 \qquad (1.2.31a)$$

Equations (1.2.30) and (1.2.31) can then be summarized as follows:

▶ A given beam of photons is in a pure polarization state if and only if $P = 1$. If $P < 1$ the beam is in a mixed state.

If a beam is such that $P > 0$ we will refer to the beam as "*polarized*" (*completely polarized* if $P = 1$); if $P = 0$ we will refer to the beam as "*unpolarized.*" In the lattter case all Stokes parameters vanish and the corresponding density matrix is given by

$$\rho = \frac{I}{2}\begin{pmatrix} 1 & 0 \\ 0 & 1 \end{pmatrix} \qquad (1.2.32)$$

Since the Stokes parameters vanish in *any* representation when $P = 0$ Eq. (1.2.33) is independent of the choice of the basis states. Any mixture of independently prepared states $|e_1\rangle$ and $|e_2\rangle$ of opposite polarization (for example, $|+1\rangle$ and $|-1\rangle$ or $|e_x\rangle$ and $|e_y\rangle$) and equal intensities $I_1 = I_2 = I/2$ is represented by the density matrix (1.2.33). All these mixtures behave identically in their polarization properties and can be used as models for unpolarized light.

1.2.5.4. "Operational" Definition of ρ

At this point we will invert some of the results given above, as follows. In order to determine the polarization properties of a given light beam it is necessary to perform four independent measurements by, for convenience, determining the Stokes parameters. These four parameters then serve as data which enable the density matrix to be *defined* by Eq. (1.2.24). The result of any further experiment performed on the beam can then be calculated with the help of Eq. (1.2.19) or (1.2.29).

A beam of photons is in a pure polarization state if and only if $P = 1$ (completely polarized beam). In this case the polarization state of the beam can be represented by a single-state vector $|e\rangle$. In this case the Stokes parameters are not independent; because of condition (1.2.30a) three of the parameters suffice for a complete characterization of the beam [two parameters in the case of the normalization (1.2.15)].

Finally, a beam of photons is in a mixed state if $P < 1$. In the special case $P = 0$ the beam is unpolarized and represented by the density matrix (1.2.33).

2

General Density Matrix Theory

2.1. Pure and Mixed Quantum Mechanical States

In this chapter the concepts introduced in Chapter 1 will be generalized to systems with more than two degrees of freedom. The examples discussed in the preceding sections will provide the physical background for the general treatment in this chapter. We will begin with a further discussion of pure and mixed states.

In classical mechanics the dynamical state of a system, for example, of structureless particles, is completely determined once the values of all positions and momenta of the particles are known. The state of the system at any subsequent time can then be predicted with certainty. But often only averages of the positions and momenta of the particles are given. Because of this incomplete information the methods of statistical mechanics must be applied. We are concerned here with quantum mechanical systems for which the maximum possible information is not available. However, the phrase "maximum possible information" has in quantum mechanics a more restricted meaning than in classical physics since not all physical observables can be measured simultaneously with precision. Our first task, therefore, is to discuss the meaning of the phrase "maximum information" in quantum mechanics.

As is well known, a precise simultaneous measurement of two physical variables is only possible if the two operators corresponding to the two variables commute. Thus, if two operators Q_1, Q_2 commute it is possible to find states in which Q_1 and Q_2 have definite eigenvalues q_1, q_2. Similarly, if a third operator commutes with both Q_1 and Q_2 then states can be found in which Q_1, Q_2, Q_3 have simultaneously definite eigenvalues q_1, q_2, q_3, and so on. The eigenvalues $q_1, q_2, q_3, \ldots$ can thus be used to give an increasingly precise classification of the system. The largest set of mutually commuting independent observables $Q_1, Q_2, \ldots$ that can be found will give the most complete characterization possible. (An important example is the classification of states in terms of constants of the motion.) The measurement of another variable, corresponding to an operator which does not commute with the set $Q_1, Q_2, \ldots$, necessarily introduces uncertainty into at least one of those already measured. It is therefore not possible to give a more complete specification of the system.

Thus, in general, the maximum information which can be obtained on a system (in the quantum mechanical sense) consists of the eigenvalues $q_1, q_2, \ldots$ of a complete commuting set of observables which have been measured ("complete experiment"). Once a complete experiment has been performed one can be sure that the state of the system is precisely the corresponding eigenstate of the set $Q_1, Q_2, \ldots$ associated with the measured eigenvalues $q_1, q_2, \ldots$. The system is then completely specified by assigning the state vector $|q_1, q_2, \ldots\rangle$ to it. If the measurement of the observables $Q_1, Q_2, \ldots$ on the state $|q_1, q_2, \ldots\rangle$ is immediately repeated, one can be sure to find the same values $q_1, q_2, \ldots$ again.

▶ The existence of such a set of experiments (for which the results can be predicted with certainty) gives a necessary and sufficient characterization for a state of "maximum knowledge" (Fano, 1957). *States of maximum knowledge are called pure states.*

Pure states represent the ultimate limit of precise observation as permitted by the uncertainty principle and are the quantum mechanical analog of such classical states where all positions and momenta of all particles are known.

As shown in quantum mechanics, the question when a system of commuting operators is complete can only be answered by experiment.

A complete experiment can be designed to act as a *filter* which can be used to "prepare" a system in a pure state. For example, for a beam of free electrons a complete set of commuting operators is provided by the momentum operator and the z component S_z of the spin operator. By sending a beam of electrons through a series combination of two (ideal) filters the first one selecting particles with sharp momentum $\mathbf{p}$, the second one selecting particles with sharp eigenvalue m of S_z, a beam can be prepared in the state $|\mathbf{p}, m\rangle$. That is, the particles in the beam transmitted by both filters will have the same values of $\mathbf{p}$ and m. This can be tested by sending the emerging beam through a second set of filters identical to the first pair, and it will be found that the beam will be transmitted completely. We can repeat this experiment again and again; we will always find the same values $\mathbf{p}$ and m and we can predict this result with certainty.

If only the spin properties of the beam are of interest the dependence of the state on all other variables than spin can be suppressed (for example, by considering beams where all particles have the same momentum) and the state vector can simply be denoted by $|m\rangle$ as we did in Chapter 1.

The choice of a complete, commuting set of operators is not unique. For example, instead of expanding a pure spin state in terms of the eigenstates $|p, m\rangle$ of the momentum operator and $\hat{S}_z$ the eigenstates $|p, m'\rangle$ of the momentum operator and $\hat{S}_{z'}$ can be used where z and z' are not the same. Let us consider two sets of observables $Q_1, Q_2, \ldots$ with eigenstates $|\psi\rangle = |q_1, q_2, \ldots\rangle$ and $Q'_1, Q'_2, \ldots$ with eigenstates $|\phi\rangle = |q'_1, q'_2, \ldots\rangle$, where at least one of the operators Q'_i does not commute with the first set. If a given system is represented by the state vector $|\psi\rangle$ it can always be written as a linear superposition of all eigenstates of the operators $Q'_1, Q'_2, \ldots$:

$$|\psi\rangle = \sum_n a_n |\phi_n\rangle \qquad (2.1.1)$$

where the index n distinguishes the different eigenstates. Equation (2.1.1) is the mathematical expression for the *principle of superposition*.

The particular states $|\phi_n\rangle$ that have been used in the expansion (2.1.1) are termed "basis states" and the state $|\psi\rangle$ is said to be written in the $\{|\phi_n\rangle\}$ *representation*. We will always assume that the basis states are orthonormal:

$$\langle \phi_n | \phi_m \rangle = \delta_{nm} \qquad (2.1.2a)$$

and complete:

$$\sum_n |\phi_n\rangle\langle\phi_n| = 1 \qquad (2.1.2b)$$

A direct consequence of the property (2.1.2a) is that the expansion coefficients a_n are given by

$$a_n = \langle \phi_n | \psi \rangle \qquad (2.1.3)$$

We will normalize according to

$$\langle \psi | \psi \rangle = \sum_n |a_n|^2 = 1 \qquad (2.1.4)$$

where Eq. (2.1.2a) has been used together with the expansion

$$\langle \psi | = \sum_n a_n^* \langle \phi_n | \qquad (2.1.5)$$

for the adjoint state $\langle \psi |$.

We recall that the absolute squares $|a_n|^2$ give the probabilities that a measurement will find the system in the nth eigenstate.

From Eq. (2.1.1) it follows that a pure state can be characterized in two ways. Either it can be specified by giving all eigenvalues $q_1, q_2, \ldots$ of a complete operator set, or it can be specified by the amplitudes a_n which give $|\psi\rangle$ in terms of the eigenstates $|\psi_n\rangle$ of another set of observables. The second set is usually more convenient.

In practice, a complete preparation of a system is seldom achieved, and in most cases the dynamical variables measured during the preparation do not constitute a complete set. As a result the state of the system is not pure and it cannot be represented by a single-state vector. It can be described by stating that the system has certain probabilities W_1, W_2, ... of being in the pure states $|\phi_n\rangle$, $|\phi_2\rangle$, ..., respectively. In the case of incomplete preparations, it is therefore necessary to use a statistical description in the same sense as in classical statistical mechanics.

▶ Systems which cannot be characterized by a single-state vector are called *statistical mixtures*.

Examples have already been given in Chapter 1.

Consider an ensemble of particles in the pure state $|\psi\rangle$. If this state is not one of the eigenstates of an observable Q then measurements of the corresponding physical quantity will produce a variety of results, each of which is an eigenvalue of Q. If similar measurements were made on a very large number of particles, all of which were in the same state $|\psi\rangle$, then, in general, all the possible eigenvalues of Q would be obtained. The average of the obtained results is given by the expectation value $\langle Q \rangle$ of the observable Q, which is defined by the matrix element:

$$\langle Q \rangle = \langle \psi | Q | \psi \rangle \tag{2.1.6}$$

in the normalization (2.1.4).

In order to obtain $\langle Q \rangle$ for a mixture of states $|\psi_1\rangle$, $|\psi_2\rangle$, ... the expectation values $\langle Q_n \rangle = \langle \psi_n | Q | \psi_n \rangle$ of each of the pure state components must be calculated and then averaged by summing over all pure states multiplied by its corresponding statistical weight W_n:

$$\blacktriangleright \qquad \langle Q \rangle = \sum_n W_n \langle \psi_n | Q | \psi_n \rangle \tag{2.1.7}$$

It should be noted that statistics enter into Eq. (2.1.7) in two ways: First of all in the quantum mechanical expectation value $\langle Q_n \rangle$ and secondly in the ensemble average over these values with the weights W_n. While the first type of averaging is connected with the perturbation of the system during a measurement and is therefore inherent in the nature of quantization, the second averaging is introduced because of the lack of information as to which of several pure states the system may be in. This latter averaging closely resembles that of classical statistical mechanics and it can be conveniently performed by using the density matrix techniques which will be discussed in the following section.

2.2. The Density Matrix and Its Basic Properties

Consider a mixture of independently prepared states $|\psi_n\rangle$ ($n = 1, 2, \ldots$) with statistical weights W_n. These states need not necessarily be orthonormal to each other. The density operator describing the mixture is then defined as

$$\rho = \sum_n W_n |\psi_n\rangle\langle\psi_n| \qquad (2.2.1)$$

where the sum extends over all states present in the mixture. ρ is also referred to as *statistical operator*.

In order to express the operator (2.2.1) in matrix form a convenient set of basis states must first be chosen, say, $|\phi_1\rangle, |\phi_2\rangle, \ldots$, which fulfill the condition (2.1.1). Using the superposition principle

$$|\psi_n\rangle = \sum_{m'} a_{m'}^{(n)} |\phi_{m'}\rangle \qquad (2.2.2a)$$

and

$$\langle\psi_n| = \sum_m a_m^{(n)*} \langle\phi_m| \qquad (2.2.2b)$$

then Eq. (2.2.1) becomes

$$\rho = \sum_{nm'm} W_n a_{m'}^{(n)} a_m^{(n)*} |\phi_{m'}\rangle\langle\phi_m| \qquad (2.2.3)$$

Taking matrix elements of Eq. (2.2.3) between states $|\phi_j\rangle$ and $\langle\phi_i|$ and applying the orthonormality conditions (2.1.2a) we obtain

$$\langle\phi_i|\rho|\phi_j\rangle = \sum_n W_n a_i^{(n)} a_j^{(n)*} \qquad (2.2.4)$$

The set of all elements (2.2.4), where i and j run over all basis states over which the sum in Eq. (2.2.2) extends, gives an explicit matrix representation of the operator (2.2.1), the *density matrix*. Since the basis states $|\phi_n\rangle$ have been used we will say that the set of Eqs. (2.2.4) gives the elements of the density matrix in the $\{|\phi_n\rangle\}$ *representation*.

We will now derive and generalize some important properties of the density matrix which were first encountered in Chapter 1. First of all, from

Eq. (2.2.4) it is evident that ρ is *Hermitian*, that is, the matrix (2.2.4) satisfies the condition

▶ $$\langle \phi_i |\rho| \phi_j \rangle = \langle \phi_j |\rho| \phi_i \rangle^* \tag{2.2.5}$$

Secondly, since the probability of finding the system in the state $|\psi_n\rangle$ is W_n and since the probability that $|\psi_n\rangle$ can be found in the state $|\phi_m\rangle$ is $|a_m^{(n)}|^2$ *the probability of finding the system in the state $|\phi_m\rangle$ is given by the diagonal element*

$$\rho_{mm} = \sum_n W_n |a_m^{(n)}|^2 \tag{2.2.6}$$

This relation gives a physical interpretation of the diagonal elements of ρ. The physical importance of the off-diagonal elements will be considered in Section 2.3. Because probabilities are positive numbers it follows from Eq. (2.2.6) that

▶ $$\rho_{mm} \geq 0 \tag{2.2.6a}$$

Using the same arguments as in Chapter 1 it can be shown that the probability $W(\psi)$ of finding the system in the state $|\psi\rangle$ after a measurement is given by the matrix element:

▶ $$W(\psi) = \langle \psi |Q| \psi \rangle \tag{2.2.7}$$

in the normalization (2.1.4). This becomes evident if Eq. (2.2.1) is substituted for ρ in Eq. (2.2.7):

$$W(\psi) = \sum_n W_n |\langle \psi_n |\psi\rangle|^2$$

and the coefficients $|\langle \psi_n |\psi\rangle|^2$ are interpreted according to Eq. (2.1.3).

The trace of ρ is a constant independent of the representation. From the normalization (2.1.4) and the condition

$$\sum_n W_n = 1 \tag{2.2.8a}$$

it follows that

$$\operatorname{tr} \rho = \sum_i \rho_{ii} = \sum_a W_a \sum_n |a_i^{(n)}|^2 = 1 \tag{2.2.8}$$

The expectation value of any operator Q is given by the trace of the product of ρ and Q:

▶
$$\langle Q \rangle = \sum_{mm'} \sum_n W_n a_m^{(n)} a_m^{(n)*} \langle \phi_m | Q | \phi_{m'} \rangle$$

$$= \sum_{mm'} \langle \phi_{m'} | \rho | \phi_m \rangle \langle \phi_m | Q | \phi_{m'} \rangle$$

$$= \text{tr} \, (\rho Q) \tag{2.2.9}$$

where we first inserted Eq. (2.2.2) into Eq. (2.1.7) and then applied Eq. (2.2.4). More generally, if we drop the normalization (2.2.8) (as we did in Section 1.2.4), then $\langle Q \rangle$ is given by

$$\langle Q \rangle = \frac{\text{tr} \, (\rho Q)}{\text{tr} \, \rho} \tag{2.2.9a}$$

The relation (2.2.9) is an important result. We recall from quantum mechanics that all information on the behavior of a given system can be expressed in terms of expectation values of suitably chosen operators. Thus the basic problem is to calculate the expectation values. Since the expectation value of any operator can be obtained by use of Eq. (2.2.9) *the density matrix contains all physically significant information on the system.*

So far the density matrix has been defined by Eq. (2.2.4). In general, however, it is more convenient to consider ρ to be defined by the expression (2.2.9) in the following way. As many operators $Q_1, Q_2, \ldots$ as there are independent parameters in ρ are chosen and their expectation values $\langle Q_1 \rangle, \langle Q_2 \rangle, \ldots$ are given as initial information on the system. The corresponding density matrix can then be determined by solving the set of equations

$$\text{tr} \, (\rho Q_i) = \langle Q_i \rangle$$

Once ρ is determined in this way any further expectation value can be obtained by applying Eq. (2.2.9). For example, as discussed in Section 1.1, the density matrix of spin-$1/2$ particles can be obtained from a knowledge of the three expectation values $\langle \sigma_i \rangle$, that is, the components of the polarization vector.

This method has several advantages. First of all the definition of a mixture given by Eq. (2.2.1) is not unique for the reasons described in Section 1.1.5. Secondly, the initial information on a system is often expressed in terms of expectation values of a set of operators rather than by specifying the pure states present in the mixture. This approach has been particularly advocated by Fano (1957) and we will return to it in Chapter 4.

We will now consider the number of independent parameters which are needed to specify a given density matrix. This will depend on the number of orthogonal states over which the sum in Eqs. (2.2.2) extends. In general, this number is infinite, but it is often finite when only one particular property of the system (the spin, for instance) is of interest and the dependence on all other variables can be suppressed. In the following discussion we will consider the case in which the number of basis states in the expansion (2.2.2) is equal to N. ρ is then an N-dimensional square matrix with N^2 complex elements corresponding to $2N^2$ real parameters. The Hermiticity condition (2.2.5) restricts the number of independent real parameters to N^2, and since the trace of ρ is fixed by the normalization condition it follows *that an N-dimensional density matrix is completely specified in terms of $N^2 - 1$ real parameters* [N^2 parameters if we drop the normalization (2.2.8) as, for example, in Eq. (1.2.17)]. This number may be reduced by symmetry requirements and further reduced if the system under consideration is known to be in a pure state. We will give an explicit example of this in Section 3.5.

If a given system is in a pure state, represented by the state vector $|\psi\rangle$, the corresponding density operator is given by

$$\rho = |\psi\rangle\langle\psi| \qquad (2.2.10)$$

The density matrix can be constructed in a representation in which $|\psi\rangle$ is one of the basis states. For example, a set of orthonormal states $|\psi_1\rangle = |\psi\rangle$, $|\psi_2\rangle, \ldots$ could be chosen and then, clearly, all the elements of ρ would be zero in this representation except for the element in the first row and column. It is evident from this that

$$\operatorname{tr}(\rho^2) = (\operatorname{tr}\rho)^2 \qquad (2.2.11)$$

Consider now the inverse problem of determining whether a given density matrix describes a pure state or not. In principle the problem can always be solved by transforming the given matrix to diagonal form. If this is done and it is found that all elements of ρ vanish except one, say, the ith diagonal element, then the system is in a pure state represented by the ith basis vector. However, the diagonalization is often tedious and it will therefore be useful to derive a condition which is easier to apply.

First of all, we will prove that the relation

$$\operatorname{tr}(\rho^2) \le (\operatorname{tr}\rho)^2 \qquad (2.2.12)$$

is valid in general. Consider an arbitrary density matrix which has been transformed into diagonal form with diagonal elements W_n. Then

$$\operatorname{tr}(\rho^2) = \sum_n W_n^2 \qquad (2.2.13a)$$

and

$$(\mathrm{tr}\,\rho)^2 = \left(\sum_n W_n\right)^2 \tag{2.2.13b}$$

Because the probabilities W_n are positive numbers it immediately follows that Eq. (2.2.12) is valid for the diagonal representation. Because the numerical values of traces remain unchanged under a transformation of the basis states it follows that Eq. (2.2.12) is valid in *any* representation, and not only in a diagonal one.

Suppose now that the equality sign holds in Eq. (2.2.12). In the diagonal representation this yields the condition

$$\sum_n W_n^2 = \left(\sum_n W_n\right)^2$$

from Eqs. (2.2.13). This condition can only be satisfied if all W_n vanish except one, say, W_i. Consequently, ρ contains only one nonvanishing diagonal element in the diagonal representation and the system is in a pure state represented by the ith basis vector.

In conclusion, we have proved *that Eq. (2.2.11) is a sufficient and necessary condition that a given density matrix describes a pure state.* Some consequences of this result have been discussed in Chapter 1 and further examples will be given in Chapter 3.

Finally, let us consider the case of a *random distribution* of a complete set of states $|\phi_n\rangle$. As an example, consider an ensemble of atoms with spin S and third component M characterized by state vectors $|n, S, M\rangle$, where n collectively denotes all other variables necessary to specify the states completely. Consider the case where all atoms have the same values of n and S but where the ensemble is a mixture with respect to M such that all different spin states can be found with the same probability $W_M = 1/(2S + 1)$, represented by the density operator

$$\rho = \frac{1}{2S + 1}\sum_M |nSM\rangle\langle nSM|$$

$$= \frac{1}{2S + 1}\mathbf{1} \tag{2.2.14}$$

where $\mathbf{1}$ is the $(2S + 1)$-dimensional unit matrix in spin space. Here, we have applied the completeness relation (2.1.2b) to the spin states. Evidently, the corresponding density matrix is diagonal with equal elements $1/(2S + 1)$ in *any* representation. Generalizing the definition given in Section 1.2.5 [see Eq. (1.2.33)] we will call an atomic system *unpolarized* if it can be represented by the operator (2.2.14).

2.3. Coherence versus Incoherence

2.3.1. Elementary Theory of Quantum Beats

We will begin this section with a discussion of quantum beats. The treatment given here is oversimplified in several respects. In particular the polarization of the initial photons will be completely neglected. A general theory will be presented in Chapter 5. The discussion given in this section is intended partly as an introduction to the important concept of "coherent superposition" and partly as an introduction to the topics in Chapters 3 and 5.

Consider an ensemble of atoms all of which are in their ground state $|0\rangle$ of well-defined energy E_0 (which we will put equal to zero). The atoms may be excited to higher-lying states by photon absorption. If the excitation is caused by very short pulses of light such that the duration of the pulse Δt is much smaller than the mean lifetime of the excited atoms then the excitation can be considered to have occurred "instantaneously," say, at time $t = 0$. A light pulse of duration Δt has a bandwidth $\Delta \omega \sim 1/\Delta t$ and the photons have no well-defined energy. We will assume that the energy spread $\hbar \Delta \omega$ is greater than the energy difference $E_1 - E_2$ of two atomic levels $|\phi_1\rangle$ and $|\phi_2\rangle$ (see Figure 2.1). The energy of the excited atoms will then be undefined and we represent the state of the excited atoms by a linear superposition of both states

$$|\psi(0)\rangle = a_1|\phi(0)_1\rangle + a_2|\phi(0)_2\rangle \tag{2.3.1}$$

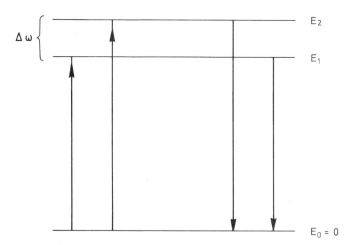

Figure 2.1. Illustration of two levels decaying to the same ground state.

immediately after the absorption (for a more detailed discussion of the underlying principles see Chapter 3).

Any state $|\phi(0)_i\rangle$ of definite energy evolves in time according to the law

$$|\phi(t)_i\rangle = \exp[-(i/\hbar)E_i t]|\phi(0)_i\rangle$$

If the decay of the excited atoms is described phenomenologically by the factor $\exp[-(1/2)\gamma_i t]$ then the time development of the state (2.3.1) is given by

$$|\psi(t)\rangle = a_1 \exp[-(i/\hbar)E_1 t - (\gamma_1/2)t]|\phi(0)_1\rangle$$
$$+ a_2 \exp[-(i/\hbar)E_2 t - (\gamma_2/2)t]|\phi(0)_2\rangle \qquad (2.3.2)$$

where γ_1 and γ_2 are the decay constants of the states $|\phi_1\rangle$ and $|\phi_2\rangle$, respectively. An expression for the intensity of the light emitted at time t can be derived from elementary radiation theory and is given by the expression

$$I(t) \sim |\langle 0|\mathbf{er}|\psi(t)\rangle|^2$$
$$= |a_1\langle 0|\mathbf{er}|\phi(t)_1\rangle + a_2\langle 0|\mathbf{er}|\phi(t)_2\rangle|^2 \qquad (2.3.3)$$

where $\mathbf{e}$ is the polarization vector of the emitted photons and $\mathbf{r}$ the dipole operator. Denoting $\langle 0|\mathbf{er}|\phi_i(0)\rangle$ by A_i and $(1/2)(\gamma_1 + \gamma_2)$ by γ and using Eq. (2.3.2) gives

$$I(t) \sim |a_1 A_1|^2 \exp(-\gamma t) + |a_2 A_2|^2 \exp(-\gamma t)$$
$$+ a_1 a_2^* A_1 A_2^* \exp[-(i/\hbar)(E_1 - E_2)t - \gamma t] \qquad (2.3.4)$$
$$+ a_1^* a_2 A_1^* A_2 \exp[+(i/\hbar)(E_1 - E_2)t - \gamma t]$$

Equation (2.3.4) shows that $I(t)$ varies periodically with a frequency $(1/\hbar)$ $(E_1 - E_2)$ (Figure 2.2). This phenomenon is known as *quantum beats* and can be understood as an *interference effect* in the sense of Eq. (2.3.3): In order to obtain $I(t)$ the amplitudes must be added before the modulus is taken. Equation (2.3.4) illustrates that it is possible to measure small energy differences by determining the beat frequency. This method is now widely used in atomic spectroscopy (see, for example, the articles in Hanle and Kleinpoppen, 1978, 1979).

2.3.2. The Concept of Coherent Superposition

It is instructive to generalize Eq. (2.3.4) as follows. The density operator of the excited atoms immediately after the excitation is given by $|\psi(0)\rangle\langle\psi(0)|$. Using the states $|\phi_1\rangle$ and $|\phi_2\rangle$ of definite energy as basis the elements of the excited state density matrix in the energy representation are found to be $\langle\phi_i|\rho(0)|\phi_j\rangle = \rho_{ij} = a_i a_j^*$ $(i, j = 1, 2)$. It is therefore plausible to

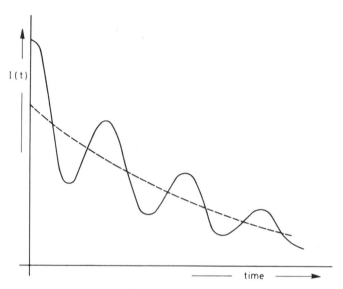

Figure 2.2. Illustration of "quantum beats."

make the following generalization of Eq. (2.3.4) (a formal proof will be given in Chapter 5). Suppose that at time $t = 0$ an excited atomic state is not pure but represented by a 2×2 density matrix $\rho(0)$ with basis vectors $|\phi_1\rangle$ and $|\phi_2\rangle$. In this case Eq. (2.3.4) still holds if the elements ρ_{ij} of this density matrix are substituted for the corresponding quantities $a_i a_j^*$. It follows *that the quantum beats are associated with the time evolution of the off-diagonal elements of the excited state density matrix* $\rho(0)$. No interference terms will occur in Eq. (2.3.4) if the density matrix $\rho(0)$ is diagonal in the energy representation.

This connection between interference and off-diagonal elements of the relevant density matrix is a general one, as will be shown by our subsequent discussions. We will therefore give the following definition. If a given system may be characterized by a density matrix written in a representation with basis vectors $|\phi_n\rangle$, then

▶ The system is a *coherent superposition* of *basis* states $|\phi_n\rangle$ if its density matrix is not diagonal in the $\{|\phi_n\rangle\}$ representation. If, in addition, the system is in a pure state it is said to be *completely coherent*. If ρ is diagonal the system is said to be an *incoherent superposition* of the *basis* states (provided there is more than one nonvanishing element) (Cohen-Tannouidji, 1962).

In this sense the time modulation of $I(t)$ is a manifestation of the coherent excitation of states with different energy as expressed by Eq. (2.3.1).

The distinction between "complete coherence" and "coherence" is often of little significance, and in the literature the term "coherence" is usually applied to both situations. We will follow this custom in cases where we are not interested whether the system under consideration is in a pure state or not. The concept of coherent superposition depends on the choice of representation for the density matrix. For example, the mixture of independently prepared states (2.2.1) is an incoherent superposition of states $|\psi_n\rangle$ but in general it is also a coherent superposition of basis states as shown by Eqs. (2.2.3) and (2.2.4).

The above definition can also be considered from the following point of view. A pure state state can always be written as a linear (completely coherent) superposition of basis states. The magnitudes and phases of the coefficients in this expansion are well defined (apart from the overall phase), that is, there exists a definite phase relationship between the basis states. The other extreme is a mixture of independently prepared basis states $|\phi_n\rangle$ represented by the density operator

$$\rho = \sum W_n |\phi_n\rangle\langle\phi_n|$$

without any definite phase relation. ρ is diagonal in the $\{|\phi_n\rangle\}$ representation and, by definition, the states $|\phi_n\rangle$ overlap incoherently. A mixture of states $|\psi_1\rangle, |\psi_1\rangle, \ldots$ represented by a density matrix which is not diagonal in the $\{|\phi_n\rangle\}$ representation [see, for example, Eqs. (2.2.3) and (2.2.4)] lies between the two extreme cases of "complete coherence" and "incoherence" with respect to $|\phi_n\rangle$.

Let us now give the following, more general, definition:

▶ A system is said to be an *incoherent superposition* of states $|\psi_1\rangle$, $|\psi_2\rangle, \ldots$ if it can be represented by the density operator

$$\rho = \sum_n W_n |\psi_n\rangle\langle\psi_n|$$

When the set $|\psi_n\rangle$ is orthonormal these states can be used as basis states and this definition is then equivalent to the one given previously.

As an example consider an atomic system being in a coherent superposition of its ground state (angular momentum $J = 0$) and an excited state with $J = 1$. The elements of the density matrix describing the system will be denoted by

$$\langle J'M'|\rho|JM\rangle = \rho(J'J)_{M'M} \tag{2.3.5}$$

In an explicit matrix form ρ is given by

$$
\rho = \begin{pmatrix}
\rho(0,0)_{00} & \rho(01)_{01} & \rho(01)_{00} & \rho(01)_{0-1} \\
\rho(10)_{10} & \rho(11)_{11} & \rho(11)_{10} & \rho(11)_{1-1} \\
\rho(10)_{00} & \rho(11)_{01} & \rho(11)_{00} & \rho(11)_{0-1} \\
\rho(10)_{-10} & \rho(11)_{-11} & \rho(11)_{-10} & \rho(11)_{-1-1}
\end{pmatrix}
\tag{2.3.6}
$$

The matrix (2.3.6) is divided into four submatrices. The upper one consists of one element, the probability $\rho(00)_{00}$ of finding the system in its ground state. The elements of the square submatrix characterize the excited state. Its diagonal elements are the probabilities of finding an atom in the corresponding substate with quantum number M. Its off-diagonal elements describe the coherence between different substates. The remaining elements in the first row and first column of the matrix (2.3.6) characterize the interference between the ground and the excited states. Density matrices of the form (2.3.6) occur, for example, in optical pumping theory corresponding to transitions $J = 0 \leftrightarrow J = 1$.

2.4. Time Evolution of Statistical Mixtures

2.4.1. The Time Evolution Operator

The time development of quantum mechanical states is described by the Schrödinger equation:

$$
i\hbar \frac{\partial |\psi(t)\rangle}{\partial t} = H |\psi(t)\rangle
\tag{2.4.1}
$$

and the equation for the adjoint state is

$$
-i\hbar \frac{\partial \langle \psi(t)|}{\partial t} = \langle \psi(t)| H
\tag{2.4.1a}
$$

The Hamiltonian may depend explicitly on the time, for example, if H contains an interaction term $V(t)$ caused by an external time-varying field. However, we will assume at present that H is time independent.

In this section we will consider how the information contained in Eq. (2.4.1) can be expressed in another useful way. We will denote an eigenstate of H with energy E_n by $|\mu_n\rangle$:

$$
H |\mu_n\rangle = E_n |\mu_n\rangle
\tag{2.4.2}
$$

If a system is represented at time $t = 0$ by an eigenstate $|\mu_n\rangle$ then at a time t the system will be found in the state

$$
|\mu_n(t)\rangle = e^{-(i/\hbar)E_n t} |\mu_n\rangle
\tag{2.4.3}
$$

which is clearly a solution of Eq. (2.4.1). Thus the time development of eigenstates of the total Hamiltonian is simply obtained by multiplying $|\mu_n\rangle$ by the exponential factor $e^{-(i/\hbar)E_n t}$.

Equation (2.4.3) can be generalized in the following way. Any solution of Eq. (2.4.1) can be expanded in terms of the set of eigenstates $|\mu_n\rangle$:

$$|\psi(t)\rangle = \sum_n C_n |\mu_n(t)\rangle$$

$$= \sum_n C_n e^{-(i/\hbar)E_n t} |\mu_n\rangle \qquad (2.4.4)$$

where the coefficients C_n are time independent. This can be shown by inserting Eq. (2.4.4) into the Schrödinger equation and applying Eq. (2.4.2). In particular, at time $t = 0$

$$|\psi(0)\rangle = \sum_n C_n |\mu_n\rangle \qquad (2.4.4a)$$

In this case (that is, where the eigenstates of a time-independent total Hamiltonian have been used as basis set) the coefficients C_n can be determined by specifying the initial conditions. Equation (2.4.4) illustrates how any state $|\psi(t)\rangle$ evolves in time. If the eigenstates and eigenvalues of H are known, then the dynamical evolution of any state vector can be predicted. Equation (2.4.4) can be written in a more abstract form. First of all, since

$$e^{-(i/\hbar)E_n t} |\mu_n\rangle = e^{-(i/\hbar)Ht} |\mu_n\rangle \qquad (2.4.5)$$

where the exponential operator function is defined by

$$e^{-(i/\hbar)Ht} = 1 - \frac{i}{\hbar}Ht - \frac{1}{2\hbar^2}H^2t^2 - \cdots \qquad (2.4.6)$$

them by operating on $|\mu_n\rangle$ with the operator given by Eq. (2.4.6) and applying Eq. (2.4.2) to any term, Eq. (2.4.5) is obtained. Substitution of the expression (2.4.5) into Eq. (2.4.4) yields

$$|\psi(t)\rangle = e^{-(i/\hbar)Ht} \sum_n C_n |\mu_n\rangle$$

$$= e^{-(i/\hbar)Ht} |\psi(0)\rangle \qquad (2.4.7)$$

$|\psi(t)\rangle$ in the form (2.4.7) may also be obtained in a more direct manner by formally integrating Eq. (2.4.1).

The operator $e^{-(i/\hbar)Ht}$ contains all the information on the time evolution of any state $|\psi(t)\rangle$ and hence also on the dynamics of the system. If the state $|\psi(0)\rangle$ of a system at time $t = 0$ is known then the state representing the system at a later time t is obtained by operating on $|\psi(0)\rangle$ with $e^{-(i/\hbar)Ht}$. If

$|\psi(0)\rangle$ is an eigenstate $|\mu_n\rangle$ of the total Hamiltonian then Eq. (2.4.7) reduces to Eq. (2.4.3). In general, however, Eq. (2.4.7) only represents a formal solution of the Schrödinger equation. Since, in order to use this equation to obtain the time development of a state, it is necessary to know the effect of the exponential operator on $|\psi(0)\rangle$, which requires, for example, a knowledge of all eigenstates and eigenvalues of H. Nevertheless, the form (2.4.7) will prove to be very useful.

We will now consider the case where *the Hamiltonian depends explicitly on the time*. In this case the Schrödinger equation

$$i\hbar\frac{\partial|\psi(t)\rangle}{\partial t} = H(t)|\psi(t)\rangle \tag{2.4.8}$$

does not have the simple solutions (2.4.4) and (2.4.7). However, Eq. (2.4.7) may be generalized by introducing an operator $U(t)$, the *time evolution operator*, which transforms a state $|\psi(0)\rangle$ into a state $|\psi(t)\rangle$:

► $$|\psi(t)\rangle = U(t)|\psi(0)\rangle \tag{2.4.9}$$

and for the adjoint states

► $$\langle\psi(t)| = \langle\psi(0)|U(t)^\dagger \tag{2.4.9a}$$

Substitution of Eq. (2.4.9) into the Schrödinger equation (2.4.8) gives

$$i\hbar\frac{\partial|U(t)}{\partial t}|\psi(0)\rangle = H(t)U(t)|\psi(0)\rangle \tag{2.4.10a}$$

Since Eq. (2.4.10a) holds for any state $|\psi(0)\rangle$ this condition can be written as an operator equation:

$$i\hbar\frac{\partial U(t)}{\partial t} = H(t)U(t) \tag{2.4.10}$$

In order to ensure that the system is in the state $|\psi(0)\rangle$ at time $t = 0$ it is necessary to impose the initial condition

$$U(0) = \mathbf{1} \tag{2.4.11}$$

For the adjoint operator we have

$$-i\hbar\frac{\partial U(t)^\dagger}{\partial t} = U(t)^\dagger H(t) \tag{2.4.12}$$

Operating on Eq. (2.4.10) on the left by $U^\dagger$ and on Eq. (2.4.12) on the right by U and then subtracting both equations gives

$$i\hbar\left(U^\dagger\frac{\partial U}{\partial t} + \frac{\alpha U^\dagger}{\alpha t}U\right) = i\hbar\frac{\partial(U^\dagger U)}{\partial t} = 0$$

It follows that $U^{\dagger}U$ must be a constant operator and since it satisfies the initial condition (2.4.11):

$$U(0)^{\dagger}U(0) = 1 \qquad (2.4.13)$$

it must also be a unitary operator. From these conditions follows that $U^{\dagger}U$ must be, the identity operator.

Equation (2.4.9) can be interpreted by noting that

$$|\langle\phi|\psi(t)\rangle|^2 = |\langle\Phi|U(t)|\psi(0)\rangle|^2$$

is the probability of finding a system at time t in the state $|\phi\rangle$ if it was represented at $t = 0$ by $|\psi(0)\rangle$.

We can summarize the contents of this section as follows. The time evolution of a state $|\psi(t)$ can be determined either by solving the Schrödinger equation (2.4.8) or, equivalently, by determining $U(t)$ by solving Eq. (2.4.10). If H is time independent we obtain by formally integrating Eq. (2.4.10):

$$U(t) = e^{-(i/\hbar)Ht} \qquad (2.4.14)$$

where the initial condition is given by (2.4.11). In this case Eq. (2.4.9) reduces to Eq. (2.4.7). For the adjoint operator we have

$$U(t)^{\dagger} = e^{+(i/\hbar)Ht} \qquad (2.4.14a)$$

In general, however, H will have an explicit time dependence and the solution of Eq. (2.4.10) will be more complicated than Eq. (2.4.15). We will consider this problem in Section 2.4.3.

2.4.2. The Liouville Equation

Suppose that at time $t = 0$ a certain mixture is represented by the density operator

$$\rho(0) = \sum_n W_n |\psi(0)_n\rangle\langle\psi(0)_n|$$

The states $|\psi(0)_n\rangle$ vary in time according to Eq. (2.4.9) and, consequently, the density operator becomes a function of time:

$$\rho(t) = \sum_n W_n |\psi(t)_n\rangle\langle\psi(t)_n|$$

$$= \sum_n W_n U(t) |\psi(0)_n\rangle\langle\psi(0)_n|U(t)^{\dagger}$$

which gives

$$\rho(t) = U(t)\rho(0)U(t)^{\dagger} \tag{2.4.15}$$

If H is time independent then

$$\rho(t) = e^{-(i/\hbar)Ht}\rho(0)\,e^{(i/\hbar)Ht} \tag{2.4.15a}$$

Differentiating Eq. (2.4.15) with respect to t and applying Eqs. (2.4.10) and (2.4.12) yields

$$i\hbar\frac{\partial\rho(t)}{\partial t} = i\hbar\frac{\partial U(t)}{\partial t}\rho(0)U(t)^{\dagger} + i\hbar U(t)\rho(0)\frac{\partial U^{\dagger}}{\partial t}$$

$$= H(t)U(t)\rho(0)U(t)^{\dagger} - U(t)\rho(0)U(t)^{\dagger}H(t)$$

Inserting Eq. (2.4.15) we obtain

$$i\hbar\frac{\partial\rho(t)}{\partial t} = [H(t), \rho(t)] \tag{2.4.16}$$

with the commutator

$$[H(t), \rho(t)] = H(t)\rho(t) - \rho(t)H(t)$$

Thus the time development of a density operator can be determined either from Eq. (2.4.15) or, equivalently, from Eq. (2.4.16). The differential equation (2.4.16) is often called the *Liouville equation* because it assumes the same form as the equation of motion for the phase space probability distribution in classical mechanics (see, for example, Tolman, 1954).

Equations (2.2.9) and (2.4.16) are basic equations of the theory. It is the simultaneous solution of these equations which leads to equations of motion for the observables. We will give an explicit example in Section 2.5.

Let us now assume that we can write

$$H(t) = H_0 + V(t) \tag{2.4.17}$$

where H_0 is assumed to be time independent and $V(t)$ describes a time-varying external field which induces transitions between the eigenstates $|\mu_n^{(0)}\rangle$ of H_0. Using these eigenstates as a basis we write

$$|\psi(t)_n\rangle = \sum_n C(t)_n|\mu(t)_n^{(0)}\rangle$$

$$= \sum_n C(t)_n\,e^{-(i/\hbar)E_n^{(0)}t}|\mu_n^{(0)}\rangle \tag{2.4.18}$$

since the time evolution of the eigenstates of H_0 is given by Eq. (2.4.3) where the eigenvalues $E_n^{(0)}$ of H_0 have been substituted for the corresponding E_n. The time dependence treated by the external force $V(t)$ is entirely contained in the coefficients $C(t)_n$ which are time independent if $V(t) = 0$, as discussed in the preceding section.

We will now derive an expression for the time evolution of the density matrix elements in the $\{|\mu_n^{(0)}\rangle\}$ representation. The matrix elements of $H = H_0 + V(t)$ are given by the expression

$$\langle \mu_{m'}^{(0)}|H(t)|\mu_m^{(0)}\rangle = E_m^{(0)}\delta_{m'm} + \langle \mu_{m'}^{(0)}|V(t)|\mu_m^{(0)}\rangle \qquad (2.4.19)$$

If the Liouville equation is multiplied by $\langle \mu_{m'}^{(0)}|$ on the left and by $|\mu_m^{(0)}\rangle$ on the right then writing $\rho(t)_{m'm} = \langle \mu_{m'}^{(0)}|\rho(t)|\mu_m^{(0)}\rangle$ we find

$$i\hbar\frac{\partial \rho(t)_{m'm}}{\partial t} = \sum_n [E_{m'}^{(0)}\delta_{m'n}\rho(t)_{nm} + \langle \mu_{m'}^{(0)}|V(t)|\mu_n^{(0)}\rangle \rho(t)_{nm}$$

$$- \rho(t)_{m'n}E_m\delta_{nm} - \rho(t)_{m'n}\langle \mu_n^{(0)}|V(t)|\mu_m^{(0)}\rangle]$$

$$= (E_{m'}^{(0)} - E_m^{(0)})\rho(t)_{m'm} + \sum_n[\langle \mu_{m'}^{(0)}|V(t)|\mu_n^{(0)}\rangle \rho(t)_{nm}$$

$$- \rho(t)_{m'm}\langle \mu_n^{(0)}|V(t)|\mu_n^{(0)}\rangle] \qquad (2.4.20)$$

This can be written in the equivalent form

$$i\hbar\frac{\partial \rho(t)_{m'm}}{\partial t} = (E_{m'}^{(0)} - E_m^{(0)})\rho(t)_{m'm} + \langle \mu_{m'}^{(0)}|[V(t), \rho(t)]|\mu_m^{(0)}\rangle$$

$$(2.4.21)$$

which is the required result.

2.4.3. The Interaction Picture

The main topic of this section will be the (approximate) determination of the time evolution operator. With the help of the obtained expressions we will also discuss the solution of the Liouville equation.

In general, an exact solution of Eq. (2.4.10) is not possible. Often however, the interaction $V(t)$ in Eq. (2.4.17) is a small perturbation and Eq. (2.4.10) can be solved by applying the methods of time-dependent perturbation theory.

To begin with some preliminary remarks, we first of all note that a large part of the time dependence of the state vectors $|\psi(t)\rangle$ is created by H_0. This is actually quite evident from Eq. (2.4.18), which contains the rapidly varying factors $e^{-(i/\hbar)E_n^{(0)}t}$. This dependence can be explicitly removed by writing Eq. (2.4.18) in the form

$$|\psi(t)\rangle = e^{-(i/\hbar)H_0 t} \sum_n C(t)_n |\mu_n^{(0)}\rangle$$

$$= e^{-(i/\hbar)H_0 t}|\psi(t)_I\rangle \qquad (2.4.22)$$

where $e^{-(i/\hbar)H_0 t}$ is defined as in Eq. (2.4.6) and where

$$|\psi(t)_I\rangle = \sum_n C_n(t)|\mu_n^{(0)}\rangle \tag{2.4.23}$$

Substituting Eq. (2.4.22) into the Schrödinger equation (2.4.8) and assuming that Eq. (2.4.17) holds gives

$$i\hbar\frac{\partial|\psi(t)\rangle}{\partial t} = i\hbar\left(-\frac{i}{\hbar}\right)H_0\, e^{-(i/\hbar)H_0 t}\,|\psi(t)_I\rangle + i\hbar(e^{-(i/\hbar)H_0 t})\frac{\partial|\psi(t)_I\rangle}{\partial t}$$

$$= [H_0 + V(t)]e^{-(i/\hbar)H_0 t}|\psi(t)_I\rangle$$

The terms containing H_0 cancel each other and we obtain the equation of motion for the state vector $|\psi(t)_I\rangle$:

$$i\hbar\frac{\partial|\psi(t)_I\rangle}{\partial t} = V(t)_I|\psi(t)_I\rangle \tag{2.4.24}$$

where we defined

▶ $$V(t)_I = e^{(i/\hbar)H_0 t}V(t)\,e^{-(i/\hbar)H_0 t} \tag{2.4.25}$$

Equation (2.4.24) shows that $\partial|\psi(t)_I\rangle/\partial t = 0$ if $V(t) = 0$, that is, *the time dependence of $|\psi(t)_I\rangle$ is created entirely by the external potential term $V(t)$*. If $V(t)$ is a small perturbation $|\psi(t)_I\rangle$ will vary slowly with time. For this reason Eq. (2.4.24) can be solved approximately within the framework of time-dependent perturbation theory and is more amenable to practical calculation than Eq. (2.4.8).

The discussions in Sections 2.4.1 and 2.4.2 relied on the fact that the state vectors $|\psi(t)\rangle$ contained all the time dependence caused by H_0 and $V(t)$, and all the information on the time development of the system. This particular description of time evolution is called the *Schrödinger picture*. As discussed above it is often convenient to remove the rapidly varying factors due to H_0 from the states. As shown by Eq. (2.4.22) this can be achieved by applying the operator

$$U(t)_0^\dagger = e^{+(i/\hbar)H_0 t} \tag{2.4.26}$$

to all Schrödinger picture states $|\psi(t)\rangle$ to give

▶ $$|\psi(t)_I\rangle = e^{(i/\hbar)H_0 t}|\psi(t)\rangle \tag{2.4.27}$$

Simultaneously, all operators $Q(t)$ can be transformed as in Eq. (2.4.25) and new operators $Q(t)_I$ can be defined as

▶ $$Q(t)_I = e^{(i/\hbar)H_0 t}Q(t)\,e^{-(i/\hbar)H_0 t} \tag{2.4.28}$$

The inverse of these transformations are given by

▶ $$|\psi(t)\rangle = e^{-(i/\hbar)H_0 t}|\psi(t)_I\rangle \qquad (2.4.27a)$$

▶ $$Q(t) = e^{-(i/\hbar)H_0 t}Q(t)_I\, e^{(i/\hbar)H_0 t} \qquad (2.4.27b)$$

Clearly, $U(t)_0$ is unitary. The time dependence of the states $|\psi(t)_I\rangle$ is now generated by the term $V(t)$ and the time dependence of the operator $Q(t)_I$ is now due to their inherent time dependence and, in addition, to the H_0 terms. The description of time evolution in terms of the states $|\psi(t)_I\rangle$ and operators $Q(t)_I$ is called the *interaction picture.*

Since $U(0)_0 = 1$, the Schrödinger and interaction pictures are the same for $t = 0$:

$$|\psi(0)\rangle = |\psi(0)_I\rangle \qquad (2.4.28)$$

After these introductory remarks we return to the problem of determining the time evolution operator. This will be done using the interaction picture. The time evolution of a Schrödinger picture state $|\psi(t)\rangle$ is described by Eq. (2.4.9). An analogous expression can be established for their interaction picture counterparts $|\psi(t)_I\rangle$. Substituting Eq. (2.4.9) into Eq. (2.4.27) gives

$$|\psi(t)_I\rangle = e^{(i/\hbar)H_0 t}U(t)|\psi(0)\rangle$$

Applying relation (2.4.28)

$$|\psi(t)_I\rangle = U(t)_I|\psi(0)_I\rangle \qquad (2.4.29)$$

where

▶ $$U(t)_I = (e^{(i/\hbar)H_0 t})U(t) \qquad (2.4.30)$$

and the inverse relation is

$$U(t) = (e^{-(i/\hbar)H_0 t})U(t)_I \qquad (2.4.31)$$

The operator $U(t)_I$ determines the time development of the states in the interaction picture.

In order to determine $U(t)_I$ Eq. (2.4.31) is inserted into Eq. (2.4.10). The terms involving H_0 cancel and we obtain

$$i\hbar\frac{\partial U(t)_I}{\partial t} = V(t)_I U(t)_I \qquad (2.4.32)$$

Where $V(t)_I$ is the solution of this equation subject to the initial condition:

$$U(0)_I = \mathbf{1} \qquad (2.4.33)$$

Equation (2.4.32) shows that the time dependence of $U(t)_I$ is entirely due to $V(t)$. For this reason time-dependent perturbation theory can be more conveniently applied to Eq. (2.4.32) than to its counterpart (2.4.10) in the Schrödinger picture.

In order to solve Eq. (2.4.32) we first formally integrate the equation to obtain

$$U(t)_I = 1 - \frac{i}{\hbar} \int_0^t V(\tau)_I U(\tau)_I \, d\tau \qquad (2.4.34)$$

where the initial condition (2.4.31) has been used. Equation (2.4.34) is not yet a solution of Eq. (2.4.32) but is merely a transformation of Eq. (2.4.32) into an integral equation which contains the unknown $U(\tau)_I$ within the integral. Equation (2.4.34) can be solved by interaction. If $V(t) = 0$ then $U(t)_I = 1$, and if $V(t)$ is sufficiently small $U(t)_I$ will only differ slightly from 1. The operator $U(\tau)_I$ can therefore be replaced by the identity operator in the integral which gives the time evolution operator in the first-order perturbation theory:

▶ $$U(t)_I = 1 - \frac{i}{\hbar} \int_0^t V(\tau)_I \, d\tau \qquad (2.4.35)$$

Inserting this relation into Eq. (2.4.34) gives the evolution operator in second-order perturbation theory and higher-order terms can be determined by further iteration.

From Eqs. (2.4.34) and (2.4.35) the corresponding expressions for the Schrödinger picture operator $U(t)$ can be derived with the help of Eq. (2.4.30).

We now proceed to the equation of motion for the density operator. Applying the unitary transformation (2.4.26) to the density operator (2.4.15) in the Schrödinger picture gives

$$\rho(t)_I = \sum_n W_n |\psi(t)_{n,I}\rangle\langle\psi(t)_{n,I}| \qquad (2.4.36)$$

where the density operator in the interaction picture is defined as

$$\rho(t)_I = (e^{(i/\hbar)H_0 t})\rho(t)(e^{-(i/\hbar)H_0 t}) \qquad (2.4.37)$$

Substitution of Eq. (2.4.15) into Eq. (2.4.37) then yields the equation of motion:

▶ $$\rho(t)_I = U(t)_I \rho(0)_I U(t)_I^\dagger \qquad (2.4.38)$$

with

$$\rho(0) = \rho(0)_I \qquad (2.4.39)$$

Similarly, substituting

$$\rho(t) = e^{-(i/\hbar)H_0 t} \rho(t)_I \, e^{+(i/\hbar)H_0 t} \qquad (2.4.40)$$

into Eq. (2.4.16) gives the *Liouville equation in the interaction picture*:

▶
$$i\hbar \frac{\partial \rho(t)_I}{\partial t} = [V(t)_I, \rho(t)_I] \qquad (2.4.41)$$

In order to obtain an approximate solution of Eq. (2.4.41) we first transform it into an integral equation. Formal integration gives

▶
$$\rho(t)_I = \rho(0)_I - \frac{i}{\hbar} \int_0^t [V(\tau)_I, \rho(\tau)_I] \, d\tau \qquad (2.4.42)$$

This integral equation can be solved iteratively in a similar way to Eq. (2.4.34). Suppose, for example, that $V(t) = 0$ for all times $t \leq 0$. The given mixture is then represented in the interaction picture by the time-independent density operator $\rho(t)_I = \rho(0)_I$. If at all times $t > 0$ a perturbation $V(t)$ is applied, then if $V(t)$ is small for times $t > 0$, $\rho(t)_I$ will not change substantially from its initial value $\rho(0)_I$.

Thus $\rho(\tau)_I$ can be replaced in the integral by its initial value $\rho(0)_I$ and the solution of Eq. (2.4.42) in first-order perturbation theory is obtained as

$$\rho(t)_I = \rho(0)_I - \frac{i}{\hbar} \int_0^t [V(\tau)_I, \rho(0)_I] \, d\tau \qquad (2.4.43)$$

$\rho(\tau)_I$ can be iterated in this equation as before to give higher-order terms.

The relations derived in this section will be illustrated by our discussions in the following chapters. A more detailed discussion of the various "pictures" used to describe time evolution may be found in any textbook on quantum mechanics.

2.5. Spin Precession in a Magnetic Field

As an example of the use of the formalism presented in the preceding sections we will now consider the precession of spin-1/2 particles in a static magnetic field. The components of the magnetic moment of a spin-1/2 particle are given by

$$\mu_i = \tfrac{1}{2}\gamma \hbar \sigma_i \qquad (2.5.1)$$

$(i = x, y, z)$, where γ is the gyromagnetic ratio and σ_i denote the Pauli matrices. The interaction between particles possessing the magnetic moment μ_i and an external magnetic field $\mathbf{H}$ is described by the Hamiltonian:

$$H = -\boldsymbol{\mu} \cdot \mathbf{H}$$
$$= -\tfrac{1}{2}\gamma \hbar \sum_j \sigma_j H_j \qquad (j = x, y, z) \qquad (2.5.2)$$

The polarization vector will change with the time and the density matrix ρ of the particles will become time dependent. The rate of change of the polarization vector $\mathbf{P}$ is determined by Eqs. (2.2.9), (2.4.16), and (2.5.2):

$$i\hbar\frac{\partial\langle\sigma_i\rangle}{\partial t} = i\hbar\frac{\partial}{\partial t}(\text{tr}\,\rho\cdot\sigma_i)$$

$$= i\hbar\,\text{tr}\left(\frac{\partial\rho}{\partial t}\sigma_i\right)$$

$$= \text{tr}\,\{[H,\rho(t)]\sigma_i\}$$

$$= \text{tr}\,\{[\sigma_i,H]\rho(t)\}$$

$$= -\tfrac{1}{2}\gamma\hbar\sum_j H_j\,\text{tr}\,\{[\sigma_i,\sigma_j]\rho(t)\} \qquad (2.5.3a)$$

where we used that the trace is invariant under cyclic permutations of the operators.

Substitution of expression (1.1.45) for ρ into Eq. (2.5.3a) yields

$$i\hbar\frac{\partial\langle\sigma_i\rangle}{\partial t} = -\tfrac{1}{4}\gamma\hbar\sum_j H_j\left(\text{tr}\,[\sigma_i,\sigma_j] + \sum_k P_k\,\text{tr}\,([\sigma_i,\sigma_k])\right) \qquad (2.5.3)$$

Application of Eq. (1.1.42a) gives

$$\text{tr}\,[\sigma_i,\sigma_j] = 0$$

and of Eq. (1.1.42b)

$$\text{tr}\,([\sigma_i,\sigma_j]\sigma_k) = 4i\varepsilon_{ijk}$$

because of the antisymmetry of the tensor ε_{ijk} [see Eq. (1.1.39)]. Inserting these results into Eq. (2.5.3) gives

$$\frac{d\langle\sigma_i\rangle}{dt} = -\gamma\sum_{jk} H_j P_k \varepsilon_{ijk} \qquad (2.5.4)$$

For example, for the component P_x from Eqs. (1.1.39) and (2.5.4),

$$\frac{dP_x}{dt} = -\gamma(H_y P_z - H_z P_y)$$

$$= +\gamma(\mathbf{P}\times\mathbf{H})_x \qquad (2.5.4a)$$

where the subscript denotes the x component of the vector product of $\mathbf{P}$ and the field $\mathbf{H}$. In vector notation Eq. (2.5.4) can be written as

$$\frac{d\mathbf{P}}{dt} = \gamma(\mathbf{P} + \mathbf{H}) \qquad (2.5.5)$$

Note that Eq. (2.5.5) is just the classical equation for the precession of the vector **P** around the direction of the field.

The derivation of Eq. (2.5.5) is a good example of the use of the basic equations (2.2.9) and (2.4.16) in deriving the equations of motion. It also illustrates the considerable ease with which this calculation can be performed through the use of the compact expression (1.1.45) for the density matrix and the subsequent use of the algebraic properties of the Pauli matrices expressed by Eq. (1.1.38). Equation (2.5.5) can also be derived without using density matrix techniques but the calculation is then considerably more tedious.

2.6. Systems in Thermal Equilibrium

A very important application of the density matrix is the one to a dynamical system which is in thermal equilibrium with the surrounding medium. It is shown in quantum statistical mechanics that the state of a system at temperature T can be represented by the density operator

$$\rho = \exp{(-\beta H)}/Z \qquad (2.6.1)$$

where $\beta = 1/kT$ and k is the Boltzmann constant. The partition function

$$Z = \text{tr} \exp{(-\beta H)} \qquad (2.6.2)$$

ensures that the normalization condition

$$\text{tr}\,\rho = 1 \qquad (2.6.3)$$

is satisfied. Equation (2.6.1) holds for a canonical ensemble, that is, a system with a constant volume, a constant number of particles, and a given mean value $\langle H \rangle$ of the Hamiltonian.

The density operator (2.6.1) plays the same role in quantum statistics as the canonical distribution function in classical statistical mechanics. This equivalence can be shown by considering the energy representation, $H|n\rangle = E_n|n\rangle$ in which the density matrix elements are given by

$$\langle n'|\rho|n \rangle = [\exp{(-\beta E_n)}/Z]\delta_{n'n} \qquad (2.6.4)$$

The diagonal elements $\langle n|\rho|n\rangle$ give the probability of finding the system in the state with energy E_n, respectively. Hence a system in thermal equilibrium is represented by an incoherent sum of energy eigenstates n with statistical weights proportional to the Boltzmann factor $\exp{(-\beta E_n)}$.

The expectation values $\langle Q \rangle$ of an operator Q acting on the system is given by

$$Q = (1/Z)\,\text{tr}\,[Q\exp{(-\beta H)}] \qquad (2.6.5)$$

which follows from Eq. (2.2.9).

Equations (2.6.1) and (2.6.4) will be applied in Chapter 7. Here we will illustrate these equations with a simple example. Consider a system of spin-1/2 particles subjected to a static magnetic fields H_z in the z direction. The Hamiltonian is given by Eq. (2.5.2):

$$H = -\boldsymbol{\mu} \cdot \mathbf{H}$$

$$= -\gamma\hbar H_z \sigma_z /2$$

The macroscopic magnetization $\mathbf{M}$ of the system is defined as

$$M_i = N\gamma\hbar\langle\sigma_i\rangle/2 \qquad (2.6.6)$$

where N is the number of particles per unit volume. Under thermal equilibrium the density matrix is diagonal and the magnetic substates will be populated according to the Boltzmann distribution (2.6.4). Hence $M_x = M_y = 0$ and

$$M_z = N\gamma\hbar\,\text{tr}\,[\exp{(-\beta H)}\sigma_z]/2Z \qquad (2.6.7)$$

Suppose that the temperature is sufficiently high to justify setting $\exp{(-\beta H)} \approx 1 - \beta H = 1 + \beta\gamma\hbar H_z \sigma_z /2$ and using $\text{tr}\,\sigma_z = 0$ and $\text{tr}\,\sigma_z^2 = 2$ we obtain

$$M_z = N\beta\gamma^2\hbar^2 H_z/2Z$$

From Eq. (2.6.2)

$$Z = \sum_m \langle m|\exp{(-\beta H)}|m\rangle$$

$$\approx 2 \qquad (2.6.8)$$

in the high-temperature limit, and finally

$$M_z = N\beta\gamma^2\hbar^2 H_z/4 \qquad (2.6.9)$$

which is the Curie law for the magnetization of spin-1/2 particles.

3

Coupled Systems

3.1. The Nonseparability of Quantum Systems after an Interaction

In Chapter 2 the basic equations of motion, Eq. (2.4.15) and the Liouville equation (2.4.16), were derived and applied to the description of interactions between a quantum system and an external *classical* field. In this chapter we will consider the problem of describing the state of a quantum system which is interacting with other (detected or undetected) *quantum* systems. This is an important topic and is of central importance to the following discussions in this book. The quantum mechanical theorems which will be introduced provide the basis for the understanding of phenomena such as quantum beats, angular correlations, and spin-depolarization effects. In Sections 3.1 and 3.2 we will describe the general theory which is required and in Sections 3.3, 3.4, and 3.5 we will give specific examples which illustrate the meaning and application of the theorems.

To begin with consider the following situation. Two separated, noninteracting systems of particles are brought together and allowed to interact with each other. We will consider the problem of analyzing the final state of the system when the constituent systems are again separated and have ceased to interact. An example of the kind of process for which this would be relevant would be the scattering of two particle beams, for example, electrons and atoms. We will denote the two subsystems by the symbols Φ and φ, respectively. A complete set of orthogonal state vectors $|\phi_i\rangle$ can be chosen to describe the Φ system, so that any state of the system can be written as a linear superposition of the basis states $|\Phi_i\rangle$. Similarly, a set of basis states $|\varphi_j\rangle$ can be chosen to describe the φ system. The indices i and j refer to the set of quantum numbers which are necessary to completely specify each system. If before the interaction the two separated systems were in *pure* states represented by the vectors $|\Phi_\alpha\rangle$ and $|\varphi_\beta\rangle$, respectively (these need not necessarily be members of the chosen basis sets), then the state of the combined systems prior to the interaction is represented by a well-defined state vector $|\psi_{in}\rangle = |\Phi_\alpha\rangle|\varphi_\beta\rangle$ in the composite space (see Appendix A).

During the interaction the time development of the state vector $|\psi_{in}\rangle$ is determined by the relevant time evolution operator which is a *linear* operator in the *composite* space. Since linear operators transform a *single* state vector into another *single* state vector, the initial pure state $|\psi_{in}\rangle$ must evolve such that the final state of the *combined system* can also be represented by a single state vector which will be denoted by $|\psi(\alpha\beta)_{out}\rangle$:

$$|\psi_{in}\rangle = |\Phi_\alpha\rangle|\varphi_\beta\rangle \to |\psi(\alpha\beta)_{out}\rangle \qquad (3.1.1a)$$

where the arrow symbolizes the effect of the time evolution operator. The state $|\psi(\alpha\beta)_{out}\rangle$ depends on the variables of both subsystems. This can be seen explicitly by expanding $|\psi(\alpha\beta)_{out}\rangle$ in terms of basis states $|\Phi_i\rangle|\varphi_j\rangle = |\phi_i\varphi_j\rangle$ of the uncoupled systems:

$$|\psi(\alpha\beta)_{out}\rangle = \sum_{ij} a(ij, \alpha\beta)|\Phi_i\rangle|\varphi_j\rangle \qquad (3.1.1b)$$

where the sum may include integrals over continuous variables.

The coefficient $a(ij, \alpha\beta)$ is the probability amplitude for a transition $|\Phi_\alpha\rangle|\varphi_\beta\rangle \to |\Phi_i\rangle|\varphi_j\rangle$ so that the absolute square $|a(ij, \alpha\beta)|^2$ gives the probability of finding a particle of the Φ system in the state $|\Phi_i\rangle$ and simultaneously a particle of the φ system in the state $|\varphi_j\rangle$ after the interaction. Only those combinations $|\Phi_i\rangle|\varphi_j\rangle$ which are allowed by the conservation laws will contribute to the expansion (3.1.1). In other words, a particular final state $|\Phi_i\rangle$ is correlated to one (or several) final states $|\varphi_j\rangle$ in such a way that all the relevant conservation laws are fulfilled.

In general, the sum in Eq. (3.1.1b) contains more than one term. The essential point is that the amplitudes depend on the variables of *both* subsystems. Consequently, it is not possible to write Eq. (3.1.1b) in the form $|\psi_{out}\rangle = |\Phi\rangle|\varphi\rangle$, where $|\Phi\rangle$ is a state vector depending *solely* on the variables of the Φ system and $|\varphi\rangle$ is a state vector depending *solely* on the variables of the φ system. In fact, such a separation of this kind would destroy the correlations which necessarily exist between the two component systems.

In order to clarify this point consider the case in which the two systems have not interacted at all. In this case the probability of finding the Φ system in a state $|\phi_i\rangle$ and the φ system in a state $|\varphi_j\rangle$ are independent of each other and the amplitudes can be factorized as $a(ij, \alpha\beta) = a(i, \alpha)a(j, \beta)$ (see Appendix A). Substituting this into Eq. (3.1.1b) then yields

$$|\psi(\alpha\beta)_{out}\rangle = \left(\sum_i a(i, \alpha)|\phi_i\rangle\right)\left(\sum_j a(j, \beta)|\varphi_j\rangle\right) \qquad (3.1.2a)$$

$$= |\Phi\rangle|\varphi\rangle \qquad (3.1.2b)$$

where $|\Phi\rangle$ and $|\varphi\rangle$ are defined by the first and second factors in (3.1.2a), respectively. In this case $|\Phi\rangle$ depends solely on the variables of the Φ system, and $|\varphi\rangle$ depends only on the variables of the φ system. In general, however, once the two systems have interacted in the past, the probability amplitudes are correlated and cannot be factorized. These results can be summarized as follows:

▶ If two systems have interacted in the past it is, in general, not possible to assign a single state vector to either of the two subsystems.

This is the essence of what is sometimes called the *principle of nonseparability* (d'Espagnat, 1976). We have shown that this principle is a direct consequence of the general rules of quantum mechanics. It should be noted that this principle has important conceptional implications and is the source of many discussions on the interpretation of quantum mechanics culminating, perhaps, in the famous Einstein–Rosen–Podolsky argument (see, for example, d'Espagnat, 1976; Jammer, 1974). An important consequence of this principle is the following. Suppose that only one of the systems, the φ system, is observed after the interaction. Although both systems were initially in pure states the interaction produces correlations between the two systems and, hence, at a later time φ will be found in a mixed state. Thus *the nonobservation of the Φ system results in a loss of coherence in the φ system.* This important result of the principle of nonseparability will be illustrated by various examples in Sections 3.3 and 3.4, where coherence only between degenerate states will be discussed. The more general case of coherently excited states with different energies, which have been excited coherently, requires a more detailed discussion of the time evolution of the system and will be considered in Chapter 5.

3.2. Interaction with an Unobserved System. The Reduced Density Matrix

Consider two (or more) interacting quantal systems. In many cases only one of the component systems is of interest and the others are left undetected. We will denote the states of the system of interest by $|\varphi_j\rangle$, the states of the undetected systems collectively by $|\Phi_i\rangle$, and the elements of the density matrix $\rho(t)$ describing the total system at time t, by $\langle\Phi_i\varphi_{j'}|\rho(t)|\Phi_i\varphi_j\rangle$. It was shown in the preceding section that, because of the interaction, the φ system is in a mixed state. It is therefore necessary to consider how the relevant density matrix $\rho(\varphi, t)$ which characterizes the system of interest alone, can be constructed.

Consider an operator $Q(\varphi)$ which acts only on the variables of the φ system, that is, its matrix elements are given by

$$\langle \Phi_{i'}\varphi_{j'}|Q(\varphi)|\Phi_i\varphi_j\rangle = \langle \varphi_{j'}|Q(\varphi)|\varphi_j\rangle\delta_{i'i} \qquad (3.2.1)$$

where orthogonality of the states $|\phi_i\rangle$ has been assumed. The expectation value $\langle Q(\varphi)\rangle$ is found using Eqs. (2.2.9) and (3.2.1):

$$\langle Q(\varphi, t) = \text{tr}\,\rho(t)Q(\varphi)$$

$$= \sum_{i'ij'j} \langle \Phi_{i'}\,\varphi_{j'}|\rho(t)|\Phi_i\varphi_j\rangle\langle \Phi_i\varphi_j|Q(\varphi)|\Phi_{i'}\varphi_{j'}\rangle$$

$$= \sum_{j'j} \left[\sum_i \langle \Phi_i\varphi_{j'}|\rho(t)|\Phi_i\varphi_j\rangle \right] \langle \varphi_j|Q(\varphi)|\varphi_{j'}\rangle \qquad (3.2.2)$$

Defining the elements of a matrix $\rho(\varphi, t)$ by

▶
$$\langle \varphi_{j'}|\rho(\varphi, t)|\varphi_j\rangle = \sum_i \langle \Phi_i\varphi_{j'}|\rho(t)|\Phi_i\varphi_j\rangle \qquad (3.2.3)$$

Eq. (3.2.2) can be written in the form

▶
$$\langle Q(\varphi, t)\rangle = \sum_{j'j} \langle \varphi_{j'}|\rho(\varphi, t)|\varphi_j\rangle\langle \varphi_j|Q(\varphi)|\varphi_{j'}\rangle$$

$$= \text{tr}\,\rho(\varphi, t)Q(\varphi) \qquad (3.2.4)$$

which is of the form of Eq. (2.2.9).

All information on the φ system can be expressed in terms of expectation values $\langle Q(\varphi, t)\rangle$ of as many operators as necessary. It follows from Eq. (3.2.4) that any of these expectation values can be calculated once $\rho(\varphi, t)$ is known. *In this sense $\rho(\varphi, t)$ contains all information on the φ system.*

$\rho(\varphi, t)$ is usually called the *reduced density matrix*. As is shown by Eq. (3.2.3) it is obtained by taking those matrix elements of the total density matrix $\rho(t)$ which are diagonal in the unobserved variable i, and summing these elements over all i. In this way all nonessential indices can be eliminated. In essence, the total density matrix $\rho(t)$ is calculated and then projected onto the subspace of interest. This is a most useful property of the density matrix and considerable use of it will be made in the rest of this book.

For convenience we will introduce the shorthand notation

$$\text{tr}_\Phi\,\rho(t) = \sum_i \langle \Phi_i|\rho(t)|\Phi_i\rangle$$

where tr_Φ is the trace over all unobserved variables. Equation (3.2.3) can then be written in the form

$$\langle \varphi_{j'} | \rho(\varphi, t) | \varphi_j \rangle = \langle \varphi_{j'} | \left[\sum_i \langle \Phi_i | \rho(t) | \Phi_i \rangle \right] | \varphi_j \rangle$$

$$= \langle \varphi_{j'} | \operatorname{tr}_\Phi \rho(t) | \varphi_j \rangle$$

or in operator notation

► $$\rho(\varphi, t) = \operatorname{tr}_\Phi \rho(t) \qquad (3.2.5)$$

A quantum mechanical system which is *closed*, or isolated from the rest of the world, has a "Hamiltonian" evolution, that is, there exists a time-independent Hamiltonian H and a unitary operator $U(t) = \exp(-iHt/\hbar)$ such that the temporal evolution of the system is given by the unitary transform (2.4.15) or, alternatively, by the Liouville equation (2.4.16). Under such an evolution a pure state is always transformed into another pure state so that mixtures can neither be created nor destroyed.

Consider now the interaction between a quantum system and external "classical" forces. The term *classical* means that the reaction of the system back on the source of the fields can be neglected. Examples are semiclassical radiation theories or the theory of potential scattering where the influence of the target on the projectile particles is approximated by a potential function. The time evolution of the quantum system is described by a unitary operator $U(t)$ and the system obeys the Liouville equation (or the Schrödinger equation in the case of a pure state). For time-dependent external forces the observables, in particular the Hamiltonian, will depend explicitly upon the time discussed in Section 2.4.

When the reaction of a quantum system (φ) back to the external world (Φ) cannot be neglected one can enlarge the φ system by Φ so that the combined system is closed and has a Hamiltonian evolution. Often Φ is undetected. In this case we will refer to φ as an *open* quantum mechanical system. The dynamical evolution of such an open system is fundamentally different from that of a closed one. In particular, as discussed in Section 3.1, the φ system will be found in a mixed state after an interaction with an unobserved quantum system even if before the interaction the component systems were in pure states. Hence

► the time evolution of an open quantum mechanical system cannot be described by the Liouville equation (or the Schrödinger equation).

In other words, the relevant reduced density matrix $\rho(\varphi, t)$ is not expressible as the unitary transform of a density matrix $\rho(\varphi, 0)$ at an earlier instant of time $t = 0$. This is an important difference between open systems and systems which are closed, or which can be described semiclassically. (This difference has played an important role, for example, in recent discussions on the validity of so-called "neoclassical" radiation theories. For details we refer particularly to the paper by Chow, Scully, and Stoner, 1975.)

The starting point for the discussion of the time variations of open systems is the Liouville equation for the corresponding "enlarged" system which includes all interacting systems. Suppose, for example, that a closed system can be divided into two interacting quantum systems φ and Φ. The combined system is described by a Hamiltonian H which consists of three parts:

$$H = H(\varphi)_0 + H(\Phi)_0 + V(\varphi, \Phi)$$

referring to the free motion of the systems and the interaction between φ and Φ, respectively. We assume that only φ is experimentally relevant and Φ undetected. Because of the interaction term V, which couples φ and Φ, no Hamiltonian exists describing the dynamics of the φ system alone. The time evolution of the reduced density matrix $\rho(\varphi, t)$ is obtained by taking the partial trace tr_Φ on both sides of Eq. (2.4.15) and using Eq. (3.2.5):

▶
$$\rho(\varphi, t) = \mathrm{tr}_\Phi\, U(t)\rho(0)U(t)^\dagger \qquad (3.2.6a)$$

or, from Eq. (2.4.16),

$$\dot{\rho}(\varphi_1 t) = -\frac{i}{\hbar}\,\mathrm{tr}_\Phi\,[H, \rho(t)] \qquad (3.2.6b)$$

where $\rho(t)$ and $U(t)$ are operators of the closed systems and H is time independent.

It should be noted that the variation in time of the combined system is *reversible* inasmuch as the initial state $\rho(0)$ can be obtained mathematically from $\rho(t)$ by the inverse transformation $\exp(iHt/\hbar) \cdot \rho(t) \exp(-iHt/\hbar)$. Open systems, however, will frequently show an *irreversible* behavior. This is a consequence of the interaction with unobserved systems (for example, a "heat bath"), which is expressed formally by the sum over all unobserved variables in Eqs. (3.2.6). The process of taking the trace provides a fundamental quantum mechanical source of irreversibility. We will discuss this in detail in Chapter 7.

The abstract results obtained in this and the preceding section deserve considerable illustration and interpretation. The remainder of this chapter will be devoted to this task where we will concentrate on some simple examples. Equations (3.2.6) will then be developed and further applied in the following chapters.

In conclusion, in this and the preceding section we have considered situations where we are confronted with coupled quantum mechanical systems only one of which (φ) is of experimental relevance. It is then economic to look for a description of φ alone. We have shown that, in general, no state vector exists which describes the dynamical behavior of a subsystem coupled with other quantum systems. *Hence an open system must be characterized in terms of its reduced density matrix.* In principle, all physical systems are interrelated since it is never possible to completely isolate a system. The conventional framework of quantum mechanics in terms of state vectors is therefore always an idealization.

We have then considered the time evolution of an open quantum mechanical system under the influence of its surroundings. The theory has to be based upon the Liouville equation which gives a complete microscopic description of closed systems. By constructing the relevant reduced density matrix, that is, by eliminating all unobserved variables, an equation describing the dynamical behavior of an open system can be obtained.

The results obtained in Sections 3.1 and 3.2 are of fundamental importance in the quantum theory of measurement. We will not discuss here this interesting but highly controversial field of modern physics. The reader is referred, for example, to d'Espagnat (1976) and the references cited therein. As an introduction we recommend Jauch (1973).

3.3. Analysis of Light Emitted by Atoms (Nuclei)

3.3.1. The Coherence Properties of the Polarization States

In order to illustrate the theory which has been developed in the preceding sections we will consider the decay of an ensemble of excited atoms (or nuclei) by photon emission. In particular, we will study the coherence which exists between states of different *polarization*.

To begin with consider an ensemble of excited atoms in identical states represented by the state vector $|\alpha_0 J_0 M_0\rangle$, where J_0 and M_0 denote the atomic angular momentum and its z component and α_0 collectively describes all other variables which are necessary for a complete specification of the states. In order to analyze the final combined state of atoms and photons we will use the procedure outlined in Section 3.1. The initial state is pure and hence it is possible to assign a single state vector $|\psi_{\text{out}}\rangle$ to the combined atom–photon system:

$$|\psi_{\text{in}}\rangle = |\alpha_0 J_0 M_0 > \ \rightarrow |\psi_{\text{out}}\rangle$$

where $|\psi_{\text{out}}\rangle$ can be expanded in terms of a set of basis states $|\alpha_1 J_1 M_1\rangle$ and $|\omega_1 \mathbf{n}_1 \lambda_1\rangle$ characterizing the final atoms and photons, respectively.

The number of possible combinations $|\alpha_1 J_1 M_1\rangle|\omega_1 \mathbf{n}_1 \lambda_1\rangle$ of final states is restricted by the relevant conservation laws (conservation of energy and angular momentum). $|\psi_{out}\rangle$ is then obtained by multiplying all the allowed combinations $|\alpha_1 J_1 M_1\rangle|\omega_1 \mathbf{n}_1 \lambda_1\rangle$ with the corresponding transition amplitudes and summing (integrating) over all discrete (continuous) variables.

The conditions which prevail in a given experiment will select a particular set of states from all the states which contribute to the expansion of $|\psi_{out}\rangle$. For the sake of simplicity we will assume that the photon detector can be tuned to accept only photons with a single frequency, say, $\omega_1 = \omega_1'$. In addition, the position of the photon detector determines the direction, $\mathbf{n}_1 = \mathbf{n}_1'$. The observation is therefore restricted to photons with sharp frequency ω_1' detected in the direction n_1'.

Since the observed photons have a fixed energy, the atoms which have emitted these photons have sharp values of α_1 and J_1 (say, α_1' and J_1') so that energy is conserved: $E(\alpha_1' J_1') = E(\alpha_0 J_0) - h\omega_1'$. Consequently, the final state vector of interest is given by the expansion

$$|\omega_{out}\rangle = \sum_{M_1 \lambda_1} a(M_1 \lambda_1, M_0)|\alpha_1' J_1' M_1\rangle|\omega_1' \mathbf{n}_1' \lambda_1\rangle \qquad (3.3.1)$$

If no polarization measurements are performed on the photons the quantum numbers M_1 and λ_1 remain undefined and Eq. (3.3.1) shows that the sum over these undetected variables must be taken. The coefficients $a(M_1 \lambda_1, M_0)$ are the probability amplitudes for the corresponding transition $|\alpha_0 J_0 M_0\rangle \to |\alpha_1' J_1' M_1\rangle|\omega_1' n_1' \lambda_1\rangle$, and the absolute square $|a(M_1 \lambda_1, M_0)|^2$ gives the probability of finding an atom in the final state $|\alpha_1' J_1' M_1\rangle$ when a photon in the state $|\omega_1' \mathbf{n}_1' \lambda_1\rangle$ has been detected. Henceforth the dependence of the states on all fixed variables will be suppressed.

It should be noted that the state vector (3.3.1) describes only a subensemble of atoms and photons, namely, only those photons registered by the detector with sharp ω_1' and $\mathbf{n}_1'$ and only those atoms which have emitted the detected photons.

Let us consider the polarization state of the photons. Equation (3.3.1) can be written as

$$|\psi_{out}\rangle = \sum_{M_1} |M_1\rangle \sum_{\lambda_1} a(M_1 \lambda_1, M_0)|\lambda_1\rangle$$

$$= \sum_{M_1} |M_1\rangle|e(M_1, M_0)\rangle \qquad (3.3.2)$$

where the state vector

$$|e(M_1 M_0)\rangle = \sum_{\lambda_1} a(M_1 \lambda_1, M_0)|\lambda_1\rangle \qquad (3.3.3)$$

describes the polarization state of the subensemble of photons emitted in a transition between states $|M_0\rangle \to |M_1\rangle$ (see Figure 3.1 for a case of dipole radiation). Equation (3.3.3) states that these photons are in identical polarization states characterized by the state vector $|e(M_1 M_0)\rangle$. Thus if the photon detector registers only those photons which have been emitted by atoms in a transition to a single state $|M_1\rangle$ then the detected photons are in the pure state $|e(M_1 M_0)\rangle$. In principle, this can be achieved by filtering the final atoms through a Stern–Gerlach filter which accepts only atoms with definite magnetic quantum number M_1. The filtered atoms can then be detected by a counter and the coincidence between this counter and the photon detector observed. The subensemble of photons which are observed in this way is necessarily completely polarized in the sense discussed in Section 1.2: The degree of polarization P has its maximum possible value $P = 1$. The actual polarization state is specified by magnitude and relative phase of the two coefficients $a(M_1 \lambda_1, M_0)$ with $\lambda_1 = +1$ and $\lambda_1 = -1$ respectively [for example, if $a(M_1, \lambda_1 = +1, M_0) = -a(M_1, \lambda = -1, M_0)$ the photons are linearly polarized in the x direction as in Eq. (1.2.12a)].

It is important to note that, in general, the detection of completely polarized photons requires that the observation is restricted to a subensemble of photons only. One exception to this is the case where the final atoms have angular momentum $J_1 = 0$. Denoting the corresponding state by $|0\rangle$ Eq. (3.3.1) can be written in the form

$$|\psi_{\text{out}}\rangle = |0\rangle \left[\sum_{\lambda_1} a(\lambda_1, M_0) |\lambda_1\rangle \right] \qquad (3.3.4)$$

Since all atoms are in the same final state the corresponding state vector can be separated out to the front of the sum in Eq. (3.3.4). The photon state is then a pure state represented by the bracket in Eq. (3.3.4), and hence in

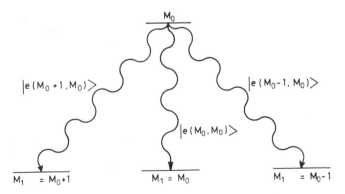

Figure 3.1. See text for explanations.

this case *all* photons emitted in the direction $\mathbf{n}'_1$ are necessarily completely polarized.

3.3.2. Description of the Emitted Photon

Let us consider the case where emitted photons are detected in direction $\mathbf{n}'_1$ (frequency ω'_1) with the final atoms unobserved.

The principle of nonseparability requires then that, in general, the state of the detected photons cannot be characterized by a single-state vector. Thus the detected radiation is not in a pure polarization state and is necessarily incompletely polarized in the sense that P is less than 1 (see Section 1.2.5). This can be demonstrated by constructing the reduced density matrix $\rho(\gamma)$ describing the photon-only system. Since the final state of the total system is represented by the state vector $|\psi_{\text{out}}\rangle$ the corresponding density matrix is simply

$$\rho_{\text{out}} = |\psi_{\text{out}}\rangle\langle\psi_{\text{out}}|$$

$$= \sum_{\substack{M'_1 M_1 \\ \lambda'_1 \lambda_1}} a(M'_1\lambda'_1, M_0)a(M_1\lambda_1, M_0)^*|M'_1\lambda'_1\rangle\langle M_1\lambda_1|$$

where Eq. (3.3.2) has been used. The elements of the reduced density matrix are then obtained by applying Eq. (3.2.3):

$$\langle\lambda'_1|\rho(\gamma)|\lambda_1\rangle = \sum_{M_1} \langle M_1\lambda'_1|\rho_{\text{out}}|M_1\lambda_1\rangle$$

$$= \sum_{M_1} a(M_1\lambda'_1, M_0)a(M_1\lambda_1, M_0)^* \tag{3.3.5}$$

This matrix corresponds to a density operator given by

$$\rho(\gamma) = \sum_{M_1\lambda'_1\lambda_1} a(M_1\lambda'_1, M_0)a(M_1\lambda_1, M_0)^*|\lambda'_1\rangle\langle\lambda_1|$$

$$= \sum_{M_1} \left[\sum_{\lambda'_1} a(M_1\lambda'_1, M_0)|\lambda'_1\rangle\right]\left[\sum_{\lambda_1} a(M_1\lambda_1, M_0)^*\langle\lambda_1|\right]$$

$$= \sum_{M_1} |e(M_1M_0)\rangle\langle e(M_1M_0)| \tag{3.3.6}$$

Equation (3.3.6) suggests the following interpretation of the operator $\rho(\gamma)$: Photons in different polarization states $|e(M_1M_0)\rangle$ can be thought of as being emitted *independently*, so that no definite phase relation exists between photons emitted in transitions to different atomic states. Thus in accordance with the definition given in Section 2.3.2, *the photon-only system can be thought of as being an incoherent superposition of states $|e(M_1M_0)\rangle$ corresponding to the various transitions.*

Since no definite phase relation exists between photons in the different polarization states these photons can in principle be distinguished (for example, by observing the final atoms in coincidence with the emitted radiation or by using suitably chosen polarization filters). The various polarization states correspond to the different "paths" by which radiation is emitted, as shown diagrammatically by the arrows in Figure 3.1. The above result is therefore often expressed in the following form: *If it is possible in principle to distinguish between photons taking the different "paths" then the total ensemble of photons can be considered as an incoherent superposition of the corresponding photon states.*

These results can be summarized in the following alternative form, due to Fano (1957):

▶   Incomplete polarization of light is necessarily associated with an incomplete determination of the final (or initial) atomic state.

It should be noted that this is a direct consequence of the principle of nonseparability. The basic result that the photon state is not pure can also be shown by proving that the matrix (3.3.5) does not satisfy condition (2.2.11).

The discussion given here is incomplete, since only the decay of excited states with a single quantum number M_0 has been considered. The important case of the excitation of atoms in superposition states with different M_0 and the decay from such superposition states will be considered via a particular example in Section 3.4.2.

3.4. Some Further Consequences of the Principle of Nonseparability

3.4.1. Collisional Spin Depolarization

As a further illustration of the theory presented in Section 3.1 we will consider elastic scattering of electrons by spin-$1/2$ atoms (or of protons or neutrons on nuclei). It will be assumed that initially both the atoms and electrons are completely polarized, for example, in states with definite values of their respective spin components, M_0 and m_0. All atoms will be assumed to be in their ground state with orbital angular momentum 0, and assumed to be sufficiently heavy to allow their recoil to be neglected. The collision will be described in the rest frame of the atoms. The electrons will be assumed to have been prepared in states with the same momentum $\mathbf{p}_0$. The atomic and electronic states can then be denoted by $|M_0\rangle$ and $|m_0\rangle$, respectively. The combined system is represented by the vector $|\psi_{in}\rangle = |M_0\rangle|m_0\rangle$ and, since the initial states are pure, the final state of the combined system can be represented by a single-state vector $|\psi_{out}\rangle$. We expand $|\psi_{out}\rangle$ in terms of state vectors $|M_1\rangle$ and $|p_1m_1\rangle$ describing the final atomic and electronic states, respectively. For simplicity it will be assumed that the electron detector is tuned to accept only electrons with fixed momentum $\mathbf{p}_1$ ($|\mathbf{p}_1| = |\mathbf{p}_0|$). The state vector of interest is then given by

$$|\psi_{out}\rangle = \sum_{M_1m_1} a(M_1m_1, M_0m_0)|M_1\rangle|m_1\rangle$$

$$= \sum_{M_1} |M_1\rangle|\chi(M_1)\rangle \tag{3.4.1}$$

where $|m_1\rangle$ implicitly describes the momentum state of the electrons as well as their spins. The two state vectors $|\chi(M_1)\rangle$ $(M_1 = \pm 1/2)$ in Eq. (3.4.1) are given by

$$|\chi(M_1)\rangle = \sum_{m_1} a(M_1m_1, M_0m_0)|m_1\rangle \tag{3.4.2}$$

and describe the state of electrons with fixed momentum $\mathbf{p}_1$ scattered by atoms which have simultaneously made the transition $|M_0\rangle \rightarrow |M_1\rangle$. In order to select a subensemble of the electrons in one of the states $|\chi(M_1)\rangle$ (say, with $M_1 = +1/2$) it must be ensured that the electron detector registers a scattered electron only when it has interacted with an atom which has undergone the transition $|M_0\rangle \rightarrow |M_1\rangle$. (See the discussion in Section 3.3.1.) Since these electrons are in a pure state they are necessarily completely polarized as demonstrated in Section 1.1. The direction of the new polarization vector with respect to the initial direction depends on the magnitudes and relative phase of the amplitudes $a(M_1m_1, M_0, m_0)$, which in turn depends on the dynamics of the scattering process.

The situation is particularly simple in the case of scattering by a spinless target. If the corresponding atomic state is denoted by $|0\rangle$ then this state will remain unchanged during the collision and it can be placed in front of the sum in Eq. (3.4.1):

$$|\psi_{out}\rangle = |0\rangle\left[\sum_{m_1} a(m_1, m_0)|m_1\rangle\right] = |0\rangle|\chi\rangle \tag{3.4.3}$$

where $|\chi\rangle$ is defined by the bracket in Eq. (3.4.3). Evidently, as shown by Eq. (3.4.3), the electrons are in a pure polarization state. Consequently, in elastic scattering processes on a spinless target the electrons observed in a fixed direction with fixed energy will be completely polarized if the initial state was pure. No depolarization is possible and only the direction of the polarization vector will change.

In the general case (3.4.1) the spin state of the detected electrons is not pure but a mixture of the two states $|\chi(M_1)\rangle$ if the final atoms are not observed. Consequently, the observed electrons are necessarily depolarized: $|P| < 1$. This can be shown by using the method described in Section 3.3. The reduced density matrix $\rho(e)$ of the detected electrons is given by

$$\langle m_1' | \rho(e) | m_1 \rangle = \sum_{M_1} a(M_1 m_1', M_0 m_0) a(M_1 m_1, M_0 m_0)^* \qquad (3.4.4)$$

With this we may associate the density operator

$$\rho(e) = \sum_{M_1 m_1' m_1} a(M_1 m_1', M_0 m_0) a(M_1 m_1, M_0 m_0)^* |m_1'\rangle\langle m_1|$$

$$= \sum_{M_1} |\chi(M_1)\rangle\langle\chi(M_1)| \qquad (3.4.5)$$

It follows that the electronic state can be thought of as being an incoherent superposition of the states $|\chi(+1/2)\rangle$ and $|\chi(-1/2)\rangle$. *The depolarization of the electrons (which were originally in a pure state) is therefore necessarily associated with an incomplete determination of the atomic state after (or before) the collision.* (For a more complete treatment see Section 3.5.1.)

3.4.2. "Complete Coherence" in Atomic Excitation

As a second example we will consider the excitation of helium atoms from their ground state $|0\rangle$ to the ^1P state by electron impact. Spin–orbit interaction can be neglected in this case, and because the initial and final atoms are spinless the transition amplitudes are spin independent (this will be shown for the general case in Section 3.5). It follows that the electronic spin has no influence on the excitation process and can be neglected. Assuming that the initial electrons have definite momentum $\mathbf{p}_0$ we separate orbital and spin part of the initial state vector and write $|\mathbf{p}_0 m_0\rangle = |\mathbf{p}_0\rangle|m_0\rangle$ and, if the spin state is not pure, the initial density matrix ρ:

$$\rho = |\mathbf{p}_0\rangle\langle\mathbf{p}_0|\rho_{\text{spin}} \qquad (3.4.6)$$

and neglect the spin components.

The relevant initial state vector is then given by $|\psi_{\text{in}}\rangle = |0\rangle|\mathbf{p}_0\rangle$. Detecting scattered electrons with momentum $\mathbf{P}_1$ the final state vector of interest is given by

$$|\psi\rangle = \left[\sum_M a(M\mathbf{P}_1, \mathbf{P}_0)|M\rangle\right]|\mathbf{P}_1\rangle$$

$$= |\psi(\mathbf{P}_1)\rangle|\mathbf{P}_1\rangle \qquad (3.4.7)$$

where M denotes the final atomic state of magnetic quantum number M. Equation (3.4.7) shows that it is possible to select an ensemble of atoms which are in *identical states* $|\psi(P_1)\rangle$ defined by the brackets in Eq. (3.4.7), by restricting the observation to electrons with fixed momentum P_1.

Let us now consider the decay of the excited atoms to the ground state by photon emission with the assumption that excitation and decay can be treated as independent processes. The emitted photons may be observed in a fixed direction $\mathbf{n}$. If electrons (with momentum $\mathbf{p}_1$) and photons are detected *in coincidence* then the observation is restricted to radiation emitted by those atoms only which are in one and the same state $|\psi(\mathbf{p}_1)\rangle$. The detected photons have therefore been emitted in a transition between the same pure states $|\psi(\mathbf{p}_1)\rangle \rightarrow |0\rangle$. As a result, the photons which are detected in the coincidence experiment are necessarily completely polarized.

Unlike the case discussed in Section 3.3 the excited atomic state considered here has no well-defined magnetic quantum number. This, however, is not important for our conclusions. The essential point is that the atoms before and after the excitation are in *identical* states. This guarantees that the detected photons are completely polarized. This can also be shown formally for the state given by Eq. (3.4.7) as follows. The radiative decay from a substate $|M\rangle$ to the ground state $|0\rangle$ is described in a way similar to Eq. (3.3.4):

$$|M\rangle \rightarrow |0\rangle \sum_{\lambda} a(\lambda, M)|\lambda\rangle \qquad (3.4.8)$$

suppressing the dependence of the photon state on the direction $\mathbf{n}$. In order to obtain the state of the photons emitted in the transition $|\psi(\mathbf{p}_1)\rangle \rightarrow |0\rangle$ Eq. (3.4.8) must be multiplied by the amplitude $a(M, \mathbf{p}_1, p_0)$ and summed over all M as in Eq. (3.4.7):

$$|\psi_{\text{out}}\rangle = |0\rangle \sum_{M\lambda} a(M\mathbf{p}_1, \mathbf{p}_0)a(\lambda, M)|\lambda\rangle$$

$$= |0\rangle \sum_{M} |e(M)\rangle \qquad (3.4.9)$$

where

$$|e(M)\rangle = \sum_{\lambda} a(M\mathbf{p}_1, \mathbf{p}_0)a(\lambda, M)|\lambda\rangle \qquad (3.4.10)$$

and $M = \pm 1, 0$. In Eq. (3.4.10) the state vectors $|e(M)\rangle$ denote the polarization state of photons emitted in a transition $|M\rangle \rightarrow |0\rangle$ (see Figure 3.2) and the coefficients $a(M, \mathbf{p}_1, \mathbf{p}_0)$ are the probability amplitudes of finding an atom in the state $|M\rangle$ when the atomic system is in the state characterized by $|\psi(\mathbf{p}_1)\rangle$. Equation (3.4.9) explicitly shows that the total beam of photons detected in a given direction is in a *pure* polarization state

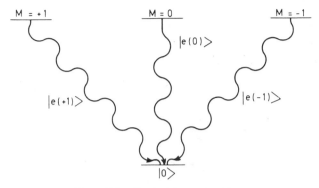

Figure 3.2. See text for explanations.

represented by the state vector

$$|e\rangle = \sum_M |e(M)\rangle$$

The main result of this section can be summarized as follows: *The complete coherence between the initial states $|M\rangle$ implies that the photon state is pure* and can therefore be represented by a completely coherent superposition of the states $|e(M)\rangle$ corresponding to the different transitions shown in Figure 3.2. The particular case which we have considered is an example of what is called a *transfer of coherence*. We will consider the problem from a more general point of view in Chapters 5 and 6.

3.5. Excitation of Atoms by Electron Impact I

3.5.1. The Reduced Density Matrix of the Atomic System

In this section we will consider the excitation of atoms by electron impact in more detail. The main assumption which will be implicit throughout this section is that all spin-dependent forces can be neglected during the collision. In particular all explicitly spin-dependent interactions between the projectile and the atoms will be neglected so that changes in the spin variables are entirely caused by *electron exchange* processes. In addition, we neglect the fine (and hyperfine) coupling inside the atom during the collision. This spin uncoupling may be understood physically as follows: In the excited atomic states the orbital angular momentum **L** and spin **S** couple under the influence of the fine-structure interaction, and precess around the total angular momentum **J** of the atom. This precession takes place in a time

$t_{LS} \sim 1/\Delta E_{LS}$, where ΔE_{LS} denotes the fine-structure splitting of the relevant level. If the collision time t_c is much shorter than the spin–orbit precession time then the spin vector will not have time to precess appreciably during the collision and $\mathbf{L}$ and $\mathbf{S}$ can be considered to be uncoupled during the collision. The state of the excited atoms immediately after the collision can then be adequately described in the LS-coupling scheme. The assumption that $t_c \ll t_{LS}$ means that the atoms can be considered as *instantaneously* excited (say, at a time $t = 0$) with respect to the much longer spin–orbit precession time (see also Chapter 5 for a more detailed discussion of this point).

Our main interest here is the description of experiments where the scattered electrons (detected in a direction $\mathbf{n}_1$) and the photons, emitted in the subsequent decay of the excited atomic states, are observed *in coincidence*. As discussed in Section 3.4.1, the observation is then restricted to radiation emitted by atoms which have been excited by the detected electrons. Thus a subensemble of atoms is "selected" in the experiment, so to speak, and this selection is the essence of the coincidence method. It is the state of this subensemble only which will be considered throughout this section.

Using the assumptions detailed above, the description of the coincidence experiment can be divided into three parts: First of all the characterization of the atomic subensemble of interest immediately after the excitation, secondly, the time evolution of the excited states under the influence of fine (and hyperfine) coupling, and finally the description of the photons observed at a time t. In this section we will commence a full treatment of the coincidence experiment with a discussion of the first part of the above program.

The atoms are assumed to be initially in their ground state with orbital angular momentum 0 and quantum numbers $\gamma_0 = \alpha_0 S_0 M_{s_0}$, where S_0 and M_{s_0} denote the atomic spin and its third component, respectively, and α_0 describes all other quantum numbers which are necessary for a complete characterization of the state. The initial state of the electrons may be characterized in terms of momentum $\mathbf{p}_0$ and spin component m_0. It will be assumed that all atoms have the same sharp values of α_0 and S_0 and the electrons the same momentum. We will use a coordinate system where the z axis is parallel to $\mathbf{p}_0$ and the x–z plane is the scattering plane ("collision system") where the scattering plane is spanned by $\mathbf{p}_0$ and $\mathbf{p}_1$.

Usually both the atoms and electrons are unpolarized in their initial states. The atomic density operator is then given by Eq. (2.2.14):

$$\rho_A = \frac{1}{2S_0 + 1} \sum_{M_{s_0}} |\alpha_0 S_0 M_{s_0}\rangle\langle\alpha_0 S_0 M_{s_0}| \tag{3.5.1a}$$

and the initial electrons are characterized in terms of the density operator:

$$\rho_e = \tfrac{1}{2} \sum_{m_0} |\mathbf{p}_0 m_0\rangle\langle \mathbf{p}_0 m_0| \tag{3.5.1b}$$

Electrons and atoms are uncorrelated before the interaction begins, and hence the density matrix ρ_{in} of the combined system factorizes and can be represented by the direct product

$$\rho_{in} = \rho_A \times \rho_e$$

$$= \frac{1}{2(2S_0 + 1)} \sum_{M_{s_0} m_0} |\alpha_0 S_0 M_{s_0} \mathbf{p}_0 m_0\rangle\langle \alpha_0 S_0 M_{s_0} \mathbf{p}_0 m_0| \tag{3.5.1c}$$

Suppressing the dependence on the fixed variables $\alpha_0 S_0 \mathbf{p}_0$ we represent the elements of ρ_{in} by

$$\langle M'_{s_0} m'_0 |\rho_{in}| M_{s_0} m_0\rangle = \frac{1}{2(2S_0 + 1)} \delta_{M_{s_0} M'_{s_0}} \delta_{m'_0 m_0} \tag{3.5.2}$$

The matrix (3.5.2) is a $2(2S_0 + 1)$-dimensional diagonal matrix in the composite spin space spanned by the $2(2S_0 + 1)$ basis states $|M_{s_0}\rangle \equiv |\alpha_0 S_0 M_{s_0}\rangle$ and $|m_0\rangle \equiv |\mathbf{p}_0 m_0\rangle$.

Using the assumption that $t_c \ll t_{LS}$ the excited atomic states immediately after the collision can be described in the LS-coupling scheme with quantum numbers $\gamma_1 = \alpha_1 L M S_1 M_{s_1}$, where M is the z component of the orbital angular momentum L. It will be assumed that only atoms in states with sharp values of $\alpha_1 L S_1$ are "selected" (experimentally, this can be achieved by resolving the emitted photons spectroscopically). The removal of this restriction will be considered in Chapter 4. In the coincidence experiments performed so far no spin analysis of the final particles has been carried out and we will therefore restrict the discussion to *the reduced density matrix characterizing the orbital states of the atomic subensemble of interest.*

For the discussion of scattering experiments it is convenient to change the normalization of the density matrix. For this purpose we will characterize a transition between states

$$\Gamma_0 = \gamma_0 \mathbf{p}_0 m_0 \rightarrow \Gamma_1 = \gamma_1 \mathbf{p}_1 m_1 \tag{3.5.3}$$

in terms of the corresponding *scattering amplitude* $f(\Gamma_1, \Gamma_0)$, which is defined as the matrix element of the transition operator T (see Appendix E for details):

$$f(\Gamma_1, \Gamma_0) = \langle \Gamma_1 |T| \Gamma_0\rangle \tag{3.5.4}$$

$f(\Gamma_1 \Gamma_0)$ will be normalized according to the condition

$$|f(\Gamma_1, \Gamma_0)|^2 = \sigma(\Gamma_1, \Gamma_0) \tag{3.5.5}$$

where $\sigma(\Gamma_1, \Gamma_0)$ is the differential cross section for the indicated transition. The equations derived in Sections 3.1 and 3.2 can be transformed to the new normalization by substituting $f(\Gamma_1, \Gamma_0)$ for the corresponding transition amplitudes $a(\Gamma_1, \Gamma_0)$ (the absolute square of which gives the probability for the corresponding transition).

Let us denote the final atomic states by $|\alpha_1 LMS_1 M_{s_1}\rangle \equiv |MM_{s_1}\rangle$ and the state of the final electrons by $|p_1 m_1\rangle \equiv |m_1\rangle$. The density matrix ρ_{out} is given by Eq. (E5):

▶
$$\rho_{\text{out}} = T\rho_{\text{in}} T^\dagger$$

Taking matrix elements between final states and applying twice the completeness relation for the initial states

$$\sum_{M_{s_0} m_0} |M_{s_0} m_0\rangle\langle M_{s_0} m_0| = 1$$

we obtain

$$\langle M'M'_{s_1} m'_1 | \rho_{\text{out}} | MMs_1 m_1\rangle$$

$$= \sum_{\substack{M'_{s_0} m'_0 \\ M_{s_0} m_0}} \langle M'M'_{s_1} m'_1 | T | M'_{s_0} m'_0\rangle\langle M'_{s_0} m'_0 | \rho_{\text{in}} | M_{s_0} m_0\rangle\langle M_{s_0} m_0 | T^\dagger | MM_{s_1} m_1\rangle$$

$$= \frac{1}{2(2S_0 + 1)} \sum_{M_{s_0} m_0} f(M'M'_{s_1} m'_1, M_{s_0} m_0) f(MM_{s_1} m_1, M_{s_0} m_0)^* \qquad (3.5.6)$$

where Eq. (3.5.2) has been used.

When no spins are observed the density matrix of interest is the reduced density $\rho(L)$ describing the orbital states of the atoms only. Using Eq. (3.2.3) the elements of this matrix can be obtained by taking the elements of ρ_{out} which are diagonal in all the unobserved variables (that is, M_{s_1} and m_1) and summing over these variables:

$$\langle M' | \rho(L) | M \rangle = \sum_{M_{s_1} m_1} \langle M'M_{s_1} m_1 | \rho_{\text{out}} | MM_{s_1} m_1\rangle$$

$$= \frac{1}{2(2S_0 + 1)} \sum_{M_{s_1} m_1 M_{s_0} m_0} f(M'M_{s_1} m_1, M_{s_0} m_0) f(MM_{s_1} m_1, m_0)^*$$

$$\equiv \langle f(M') f(M)^* \rangle \qquad (3.5.7)$$

where the notation $\langle \cdots \rangle$ indicates that the averages over the spins have been performed.

The matrix (3.5.7) is a $(2L + 1)$-dimensional matrix containing all information on the orbital system of the excited atomic subensemble of interest. In the normalization (3.5.5) the diagonal elements of ρ are given by

$$\langle M|\rho(L)|M\rangle = \frac{1}{2(2S_0 + 1)} \sum_{\substack{M_{s_1}m_1 \\ M_{s_0}m_0}} |f(MM_{s_1}m_1, M_{s_0}m_0)|^2$$

$$= \sigma(M) \qquad (3.5.8a)$$

where $\sigma(M)$ denotes the differential cross section for excitation of the magnetic substate M averaged over all spins. The trace of $\rho(L)$ gives the differential cross section σ summed over all M:

$$\text{tr}\,\rho(L) = \sum_M \sigma(M) = \sigma \qquad (3.5.8)$$

For example, the explicit expression of $\rho(L)$ for the case $L = l$ is

$$\rho(L) = \begin{pmatrix} \sigma(1) & \langle f(+1)f(0)^*\rangle & \langle f(+1)f(-1)^*\rangle \\ \langle f(+1)f(0)^*\rangle^* & \sigma(0) & \langle f(0)f(-1)^*\rangle \\ \langle f(+1)f(-1)^*\rangle^* & \langle f(0)f(-1)^*\rangle^* & \sigma(-1) \end{pmatrix} \qquad (3.5.9)$$

where the Hermiticity condition (2.2.5) has been used:

$$\langle M'|\rho(L)|M\rangle = \langle -M'|\rho(L) - M\rangle^* \qquad (3.5.10)$$

Equations (3.5.7) and (3.5.9) show that, in general, ρ has non vanishing off-diagonal elements and hence that the excited atomic subensemble of interest is a coherent superposition state of magnetic substates.

By determining the angular distribution and polarization of the emitted photons in coincidence with the scattered electrons the density matrix (3.5.9) can be completely determined (see Section 6.1). This allows more information on the scattering processes to be extracted than from traditional experiments where only the differential cross section σ is measured. In particular the off-diagonal elements of ρ contain information on the phases of the various scattering amplitudes with different M and these phases cannot be determined without the use of coincident techniques. In order to see how many measurements must be performed for complete determination of ρ the number of independent parameters specifying ρ must be determined, which will be done in the following section.

3.5.2. Restrictions due to Symmetry Requirements

In addition to the Hermiticity condition (3.5.10) the number of independent parameters describing ρ is further restricted by certain symmetry conditions. The scattering plane (x–z plane of the collision system) is defined by $\mathbf{p}_0$ and $\mathbf{p}_1$ but no direction is defined perpendicular to the scattering plane by geometry of the experiment, that is, the atomic subensemble under discussion cannot distinguish between "up" and "down" with respect to this plane. As a result, the density matrix (3.5.7) must be *invariant under reflection in the scattering plane*.

This symmetry condition is expressed by the relation

$$f(MM_{s_1}m_1, M_{s_0}m_0) = (-1)^{M+S_1-S_0}f(-M - M_{s_1} - m_1, -M_{s_0} - m_0)$$

$$(3.5.11a)$$

for the scattering amplitudes and

$$\langle M'|\rho|M\rangle = (-1)^{M'+M} < -M'|\rho| - M\rangle \qquad (3.5.11b)$$

for the density matrix. [For a proof see the textbooks on scattering theory, for example, Rodberg and Thaler (1967) and Burke and Joachain (est. 1982).] Equation (3.5.11) gives in particular

$$\sigma(M) = \sigma(-M) \qquad (3.5.12a)$$

In case $L = 1$ combining Eq. (3.5.11) with the Hermiticity condition (3.5.10) gives

$$\langle f(0)f(-1)^*\rangle = -\langle f(0)f(1)^*\rangle = -\langle f(1)f(0)^*\rangle^* \qquad (3.5.12b)$$

Furthermore, the element $\langle f(+1)f(-1)^*\rangle$ is real:

$$\langle f(+1)f(-1)^*\rangle = \langle f(-1)f(+1)^*\rangle = \langle f(-1)f(+1)^*\rangle^* \qquad (3.5.12c)$$

Thus for $L = 1$ the density matrix becomes

$$\rho(L) = \begin{pmatrix} \sigma(1) & \langle f(+1)f(0)^*\rangle & \langle f(+1)f(-1)^*\rangle \\ \langle f(+1)f(0)^*\rangle^* & \sigma(0) & -\langle f(+1)f(0)^*\rangle^* \\ \langle f(+1)f(-1)^*\rangle^* & -\langle f(+1)f(0)^*\rangle & \sigma(1) \end{pmatrix}$$

$$(3.5.13)$$

which is completely specified in terms of *five* real parameters, for example, $\sigma(1), \sigma(0), \langle f(+1)f(-1)^*\rangle$, and the real and imaginary parts of $\langle f(+1)f(0)^*\rangle$.

A convenient parametrization of the matrix (3.5.13) has been given by Hertel and Stoll (1978). In this parametrization the four parameters

$$\lambda = \frac{\sigma(0)}{\sigma}, \qquad \cos\chi = \frac{\text{Re}\,\langle f(1)f(0)^*\rangle}{[\sigma(0)\sigma(1)]^{1/2}}$$

$$(3.5.14)$$

$$\sin\Phi = \frac{\text{Im}\,\langle f(1)f(0)^*\rangle}{[\sigma(0)0(1)]^{1/2}}, \qquad \cos\alpha = \frac{\langle f(1)f(-1)^*\rangle}{\sigma_1}$$

together with the differential cross section (3.5.8) constitute a set of five independent real parameters.

The number of independent parameters can be further reduced if *spin conservation* is taken explicitly into account. Since all explicit spin-dependent terms have been neglected in the Hamiltonian describing the collision, total spin S and its z component M_s are conserved during the collision; hence

$$S = S_0 \pm \tfrac{1}{2} = S_1 + \tfrac{1}{2}, \qquad M_s = M_{s_1} + m_1 = M_{s_0} + m_0 \quad (3.5.15)$$

It is shown in textbooks on scattering theory that the dependence of the scattering amplitudes on the spin components can be factored out as

$$f(MM_{s_1}m_1, M_{s_0}m_0) = \sum_{SM_s} (S_1 M_{s_1}, \tfrac{1}{2}m_1 | SM_s)(S_0 M_{s_0}, \tfrac{1}{2}m_0 | SM_s) f(M)^{(S)}$$

$$(3.5.16)$$

where, $f(M)^{(S)}$ denotes the scattering amplitude for excitation of the magnetic substate M in the channel with total spin S. Note that the amplitudes $f(M)^{(S)}$ are independent of all spin components. The brackets in Eq. (3.5.16) denote the standard Clebsch–Gordan coefficients.

Substituting Eq. (3.5.16) into Eq. (3.5.7) and applying the orthonormality properties of the Clebsch–Gordan coefficients gives

$$\langle M' | \rho(L) | M \rangle = \langle f(M') f(M)^* \rangle \quad (3.5.17)$$

$$= \frac{1}{2(2S_0 + 1)} \sum_S (2S + 1) f(M')^{(S)} f(M)^{(S)*} \quad (3.5.18)$$

The symmetry condition (3.5.11a) reduces to

$$f(M)^{(S)} = (-1)^M f(-M)^{(S)} \quad (3.5.18a)$$

as can be shown using Eq. (3.5.16) and the symmetry properties of the Clebsch–Gordan coefficients after some algebraic manipulations. From this and Eq. (3.5.17) the following additional symmetry condition is obtained:

$$\langle M' | \rho(L) | -M \rangle = \frac{1}{2(2S_0 + 1)} \sum_S (2S + 1) f(M')^{(S)} f(-M)^{(S)}$$

$$= \frac{(-1)^M}{2(2S_0 + 1)} \sum_S (2S + 1) f(M')^{(S)} f(M)^{(S)}$$

$$= (-1)^M \langle M' | \rho | M \rangle \quad (3.5.18)$$

In the case of $L = 1$ this gives

$$\cos \alpha = -1$$

so that the density matrix (3.5.13) is completely specified by four parameters, $\sigma, \lambda, \chi, \phi$.

A smaller number of independent parameters is required if initial and final atoms are spinless ($S_0 = S_1 = 0$). In this case only one spin channel with total spin $S = 1/2$ is allowed and Eq. (3.5.17) reduces to

$$\langle M'|\rho(L)|M\rangle = f(M')f(M)^* \qquad (3.5.19)$$

No spin average is necessary in this case and the total spin S can be suppressed in this notation.

The factorization (3.5.19) of the density matrix elements into two factors, one depending only on M', the other one only on M, is typical of cases where ρ describes a pure state. In fact, it has been shown in Section 3.4.2 that, if $S_0 = S_1 = 0$, then the state of the atomic subensemble under discussion is pure and represented by the state vector

$$|\psi(\mathbf{p}_1)\rangle = \sum_M f(M)|M\rangle \qquad (3.5.20)$$

written as a completely coherent superposition of magnetic substates. In this case the atoms in the subensemble have been identically excited.

Since the amplitudes $f(M)$ satisfy the symmetry condition (3.5.18a) (with $S = 1/2$) and since the overall phase of $|\psi(P_1)\rangle$ is arbitrary the state (3.5.20) is completely specified in terms of $(2L + 1)$ parameters.

For pure states the density matrix (3.5.13) must satisfy condition (2.2.11). When this is applied to the parameters (3.5.14) this gives for the case $L = 1$

$$\cos^2 \chi + \sin^2 \phi = 1 \qquad (3.5.21)$$

and the atomic subensemble under discussion can be completely characterized in terms of the three parameters σ, λ, χ, where χ is now the relative phase between the amplitudes $f(+1)$ and $f(0)$.

In conclusion, we have considered experiments where the scattered electrons are detected in a fixed direction with momentum $\mathbf{p}_1$ and where no spin analysis of initial and final particles is performed. We have constructed the reduced density matrix $\langle M'|\rho|M\rangle$ characterizing the orbital states of the atomic subensemble excited by the detected electrons. We have in particular considered the case $L = 1$ as an example and have shown that in this case ρ is specified in terms of five independent parameters because of the Hermiticity condition and reflection invariance in the scattering plane. Using conservation of total spin this number is reduced to four. If (and only if) the atoms have been excited into *identical* states three parameters are sufficient. In this case the distinction between "coherence" (in the sense that ρ is nondiagonal) and "complete coherence" is important: In the latter case less parameters and, thus, less experiments are necessary for a complete specification of ρ.

4

Irreducible Components of the Density Matrix

4.1. Introduction

As discussed in Chapters 1 and 2 it is often useful to expand ρ in terms of a conveniently chosen operator set Q_i. This method has two main advantages. First of all, it gives a more satisfactory definition of ρ (see, for example, Section 1.1.7), and secondly by using explicitly the algebraic properties of the basis operators the calculations are often greatly simplified (see Section 2.5). The usefulness of this method depends on the choice of the basis operator set. When the angular symmetries of the ensemble of interest are important it is convenient to expand ρ in terms of irreducible tensor operators. This method provides a well-developed and efficient way of using the inherent symmetry of the system. It also enables the consequences of angular momentum conservation to be simply allowed for and enables dynamical and geometrical factors in the equation of interest to be separated from each other.

The systematic use of tensor operators was first suggested by Fano (1953). Since then they have been applied extensively, for example in angular correlation theory in nuclear physics (Steffen and Alder, 1975) and atomic physics (Blum and Kleinpoppen, 1979), in optical pumping work (Happer, 1972; Omont, 1977), descriptions of quantum beat experiments (Fano and Macek, 1973; Macek and Burns, 1976; Andrä, 1979) and experiments with laser-excited atoms (Hertel and Stoll, 1978). The material presented in this and the following chapters is directly derived from these papers.

In this chapter the theory which is central to the subsequent discussions in this book will be presented and illustrated. In Sections 4.2 and 4.3 spherical tensor operators and state multipoles will be introduced and their main properties derived. Extensive use will be made of angular momentum theory, but for the convenience of the reader some of the basic concepts are derived in the text and all the formulas which are used are listed in Appendix C. Readers who are not too familiar with the relevant mathematical techniques may omit the details of the calculations in a first reading.

The abstract theory will be illustrated by various examples. In Section 4.4 it will be shown that the state multipole description of spin systems is but a generalization of the approach of Section 1.1 which used the polarization vector. The discussion of spin-1 particles will illustrate the necessity of introducing tensors of higher rank than vectors.

In the following two sections it will be shown that the symmetry properties of a system can often be exploited more directly by using the multipole components than in terms of the density matrix elements. In Section 4.5 axially and spherically symmetric ensembles will be considered. In Section 4.6 the consequences of reflection invariance with respect to a given plane will be investigated.

In Section 4.6 we will continue with the discussion of the excited state density matrix introduced in Section 3.5 and consider another important aspect of the multipole expansion of ρ. The elements of ρ contain the full information on the scattering process; however, it is difficult to interpret these elements in terms of a physical picture and it is in this connection that the multipole components of ρ play an important role. It is often possible to anticipate the properties of the multipole components by considering the physics of the collision and we will illustrate this using the orientation vector as an example.

Finally, in Section 4.7, the time evolution of state multipoles in the presence of an internal or external perturbation will be considered.

The results obtained in this chapter will then be applied in Chapters 5 and 6 where the full power of the irreducible tensor method will become evident.

4.2. The Definition of Tensor Operators

4.2.1. The General Construction Rule

Consider two ensembles of particles, the first with angular momenta J' and the second with angular momenta J. If the two ensembles interact it is convenient to classify the possible states in terms of the total angular momentum and its z component, which we will denote here by K and Q.

Applying the usual angular momentum coupling rules gives

$$|(J'J)KQ\rangle = \sum_{M'M} (J'M', JM|KQ)|J'M'\rangle|JM\rangle \qquad (4.2.1)$$

The states $|JM\rangle$ are orthonormal since

$$\langle J'M'|JM\rangle = \delta_{M'M}\delta_{J'J} \qquad (4.2.2)$$

Let us now consider the set of operators $|J'M'\rangle\langle JM|$ defined as outer products of the angular momentum states as in Eq. (1.1.21). The state $|JM\rangle$ can be represented by a $(2J + 1)$-dimensional column vector with a unit element in the M th row and zeros elsewhere, and the corresponding adjoint state $\langle JM|$ is then represented by a row vector with a unit element in the M th column and zeros elsewhere. The outer product can then be represented by matrices using the rule (1.1.21).

It is convenient to combine the operators $|J'M'\rangle\langle JM|$ in a way similar to Eq. (4.2.1). A set of operators $T(J'J)_{KQ}$ is defined by the relation‡

$$\blacktriangleright \qquad T(J'J)_{KQ} = \sum_{M'M} (-1)^{J-M} (J'M', J-M|KQ)|J'M'\rangle\langle JM| \qquad (4.2.3)$$

The Clebsch–Gordan coefficient vanishes unless the usual angular momentum coupling rules are satisfied:

$$\blacktriangleright \qquad |J'-J| \le K \le J'+J, \qquad -K \le Q \le K \qquad (4.2.4)$$

As a consequence, for any given pair of angular momenta J' and J the number of operators (4.2.3) is limited; thus, for example, if $J' = J = 1$ the possible operators are one with $K = 0$, three with $K = 1$ (and $Q = 0, \pm 1$), and five with $K = 2$ (and $Q = \pm 2, \pm 1, 0$).

An explicit matrix representation of the operators (4.2.3) can be obtained by "sandwiching" Eq. (4.2.3) between states $\langle J'N'|$ and $|JN\rangle$ $(N' = J', \ldots, -J', N = J, \ldots, -J)$ and applying condition (4.2.2):

$$\langle J'N'|T(J'J)_{KQ}|JN\rangle$$

$$= \sum_{M'M} (-1)^{J-M} (J'M', J-M|KQ)\langle J'N'|J'M'\rangle\langle JM|JN\rangle$$

$$= (-1)^{J-N} (J'N', J-N|KQ) \qquad (4.2.5)$$

‡ We will only consider operators with integer K.

The set of all elements of a given operator $T(J'J)_{KQ}$ defines a matrix with $(2J' + 1)$ rows and $(2J + 1)$ columns:

$$T(J'J)_{KQ} =$$

$$\begin{pmatrix} \langle J'J'|T(J'J)_{KQ}|JJ\rangle & \langle J'J'|T(J'J)_{KQ}|J, J-1\rangle & \cdots & \langle J'J'|T(J'J)_{KQ}|J, -J\rangle \\ \langle J', J'-1|T(J'J)_{KQ}|JJ\rangle & \langle J', J'-1|T(J'J)_{KQ}|J, J-1\rangle & \cdots & \langle J', J'-1|T(J'J)_{KQ}|J, -J\rangle \\ \vdots & & & \\ \langle J', -J'|T(J'J)|JJ\rangle & \langle J', -J'|T(J'J)_{KQ}|J, J-1\rangle & \cdots & \langle J', -J'|T(J'J)_{KQ}|J, -J\rangle \end{pmatrix}$$

$$(4.2.6)$$

If $J' = J$ this gives a $(2J + 1)$-dimensional square matrix.

The inverse form of Eq. (4.2.3) is obtained by multiplying both sides with a Clebsch–Gordan coefficient $(J'N', J - N|KQ)$, summing over all values of K and Q, and using the orthogonality properties of the Clebsch–Gordan coefficients (see Appendix C):

$$\sum_{KQ} (J'N', J - N|KQ)T(J'J)_{KQ}$$

$$= \sum_{M'M} (-1)^{J-M}\left[\sum_{KQ} (J'N', J - N|KQ)(J'M', J - M|KQ)\right]|J'M'\rangle\langle JM|$$

$$= (-1)^{J-N}|J'N'\rangle\langle JN|$$

or

$$|J'N'\rangle\langle JN| = \sum_{KQ} (-1)^{J-N}(J'N', J - N|KQ)T(J'J)_{KQ} \qquad (4.2.7)$$

Finally, if the $3j$-symbol notation is used for the coupling coefficient the definition (4.2.3) can be rewritten as

▶ $$T(J'J)_{KQ} = \sum_{M'M} (-1)^{J'-M'}(2K + 1)^{1/2}\begin{pmatrix} J' & J & K \\ M' & -M & -Q \end{pmatrix}|J'M'\rangle\langle JM|$$

$$(4.2.8)$$

which gives

▶ $$\langle J'M'|T(J'J)_{KQ}|JM\rangle = (-1)^{J'-M'}(2K + 1)^{1/2}\begin{pmatrix} J' & J & K \\ M' & -M & -Q \end{pmatrix}$$

$$(4.2.9)$$

This particular form is of great utility since the special symmetry properties of the $3j$ symbol [Eq. (C5)] allow the corresponding symmetry properties of the tensor operators to be treated in the most direct way.

4.2.2. Transformation Properties under Rotations. The Rotation Matrix

In order to clarify the meaning of Eq. (4.2.3) we will now consider the transformation properties of the tensor operators under rotations. The angular momentum states (4.2.1) and the operators (4.2.3) are defined with respect to a fixed coordinate system, for example, with axes X, Y, Z. Suppose that a second system with axes x, y, z is obtained by the following two consecutive rotations: (i) a rotation of angle φ about the Z axis (the new axes are then x', y, Z), and (ii) a rotation of angle θ about the y direction (which transforms x' and Z into x and z, respectively). These rotations are counterclockwise when looking down the rotation axis toward the origin. The Euler angles as defined by Edmonds (1957) are then given by $\alpha = \varphi$, $\beta = \theta$, $\gamma = 0$. θ and φ are the polar angles of the z axis with respect to the XYZ system (see Figure 4.1). The x axis has polar angles

$(\theta + 90°, \varphi)$ and the y axis is specified by $(90°, \varphi + 90°)$

The angular momentum operator **J** has a component J_Z with respect to the Z axis and a component J_z with respect to the z axis. We will denote the eigenvalues of J_Z by M and the eigenvalues of J_z by m. An eigenstate $|JM\rangle$

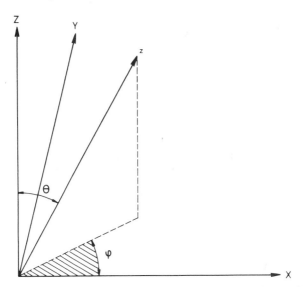

Figure 4.1.   Illustration of the rotation defined by the angles θ and φ.

of J_Z is not an eigenstate of J_z for $Z \neq z$ because in general the two operators do not commute. Using the superposition principle (2.1.1) the state $|JM\rangle$ can be written as a linear superposition of the eigenstates $|Jm\rangle$ of J_z with expansion coefficients which will depend on the angular momentum quantum numbers and the Euler angles $\omega = (\gamma, \beta, \alpha)$. It is customary to denote the expansion coefficients by $D(\omega)_{mM}^{(J)}$:

$$|JM\rangle = \sum_m |Jm\rangle D(\omega)_{mM}^{(J)} \tag{4.2.10}$$

We can interpret the coefficients as the probability amplitudes of finding a state $|Jm\rangle$ in a given state $|JM\rangle$ if the latter system is related to the former by the Euler angles ω. For fixed J the set of all coefficients can be written in form of a matrix, called *rotation matrix* the elements of which are the amplitudes $D(\omega)_{mM}^{(J)}$. Explicit expressions for various J are given, for example, in Edmonds (1957) (see also Appendix C).

The transformation law for the adjoint state $\langle JM |$ is [see Eq. (2.1.5)]

$$\langle JM | = \sum_m D(\omega)_{mM}^{(J)*} \langle Jm | \tag{4.2.11}$$

Let us now relate the operators $T(J'J)_{KQ}$, defined in the XYZ system, to operators $T(J'J)_{Kq}$ which are defined in the xyz system. In order to do this Eqs (4.2.10) and (4.2.11) are inserted into Eq. (4.2.3) and the symmetry property

$$D_{mM}^{(J)*} = (-1)^{m-M} D_{-m-M}^{(J)}$$

of the rotation matrix. Since $m - M$ is an integer $(-1)^{m-M} = (-1)^{M-m}$ and we obtain

$$
\begin{aligned}
T(J'J)_{KQ} &= \sum_{M'M} (-1)^{J'-M} (J'M', J - M | KQ) \\
&\quad \times \left[\sum_{m'} |J'm'\rangle D(\omega)_{m'M'}^{(J')} \right] \left[\sum_m D(\omega)_{mM}^{(J)*} \langle Jm | \right] \\
&= \sum_{m'm} |J'm'\rangle\langle Jm | \sum_{M'M} (J'M', J - M | KQ) \\
&\quad \times (-1)^{J-m} D(\omega)_{m'M'}^{(J')} D(\omega)_{-m-M}^{(J)} \\
&= \sum_{m'm} (-1)^{J-m} |J'm'\rangle\langle Jm | \sum_{kqq'} (J'm', J - m | kq) \\
&\quad \times \left[\sum_{M'M} (J'M', J - M | KQ)(J'M', J - M | kq') D_{qq'}^{(k)} \right] \\
&= \sum_q \left[\sum_{m'm} (-1)^{J-m} (J'm', J - m | Kq) |J'm'\rangle\langle Jm | \right] D(\omega)_{qQ}^{(K)}
\end{aligned}
\tag{4.2.12}
$$

which finally gives

$$\blacktriangleright \qquad T(J'J)_{KQ} = \sum_q T(J'J)_{Kq} D(\omega)_{qQ}^{(K)} \qquad (4.2.13)$$

In obtaining Eq. (4.2.13) the product $D(\omega)_{m'M'}^{(J')} \cdot D(\omega)_{-m-M}^{(J)}$ has been written as a linear combination of matrices $D(\omega)_{qq'}^{(k)}$ [see Eq. (C17)], performed the sum over M' and M using the orthonormality relations of the Clebsch–Gordan coefficients, and finally the definition (4.2.3) applied. Equation (4.2.13) expresses the operators $T(J'J)_{KQ}$ defined in the XYZ system, in terms of the operators $T(J'J)_{Kq}$, defined in the xyz system.

Operators which transform under rotations according to Eq. (4.2.13) are called *irreducible tensor operators of rank K and component Q*. Equation (4.2.13) shows that the rank of a tensor operator remains invariant under a rotation. We will discuss some examples of the use of this equation in the following section.

4.2.3. Examples

In this section we will consider the case of sharp angular momentum ($J' = J$) and denote the corresponding tensor operators by $T(J)_{KQ}$. First of all we will show that the operator with rank $K = 0$ is a scalar operator, that is, it remains invariant under all rotations. This can be demonstrated by showing that $T(J)_{00}$ is proportional to the $(2J + 1)$-dimensional unit matrix **1**. From the definition (4.2.8) and equation (C6) we have

$$T(J)_{00} = \sum_{M'M} (-1)^{J'-M'} \begin{pmatrix} J' & J & 0 \\ M' & -M & 0 \end{pmatrix} |JM'\rangle\langle JM|$$

$$= \frac{1}{(2J+1)^{1/2}} \sum_M |JM\rangle\langle JM|$$

$$= \frac{1}{(2J+1)^{1/2}} \mathbf{1} \qquad (4.2.14)$$

where the completeness relation $\sum_M |JM\rangle\langle JM| = 1$ has been used.

Tensor operators of rank $K = 1$ are called *vector operators*. The three vector components $T(J)_{1Q}$ can be related to the components of the angular momentum vector **J** with respect to the fixed XYZ system, J_X, J_Y, J_Z, as follows. We introduce the *spherical vector components* defined as

$$J_{\pm 1} = \mp \frac{1}{2^{1/2}} (J_X \pm iJ_Y), \qquad J_0 = J_Z \qquad (4.2.15)$$

Thus

$$J_0|JM\rangle = M|JM\rangle$$

and, for sharp J, J_0 can be represented by a $(2J + 1)$-dimensional diagonal matrix with elements

$$\langle JM'|J_0|JM\rangle = M\delta_{M'M} \qquad (4.2.16a)$$

$T(J)_{10}$ is represented by the matrix

$$\langle JM'|T(J)_{10}|JM\rangle = (-1)^{J-M'}\begin{pmatrix} J & J & 1 \\ M' & -M & 0 \end{pmatrix}$$

$$= \left[\frac{3}{(2J+1)(J+1)J}\right]^{1/2} M\delta_{M'M} \qquad (4.2.16b)$$

which satisfies Eq. (4.2.8). Similarly, using the standard results of angular momentum theory:

$$J_{\pm 1}|JM\rangle = \mp\frac{1}{2^{1/2}}[(J \mp M)(J \pm M + 1)]^{1/2}|JM \pm 1\rangle$$

the matrix representations of the other components are

$$\langle JM'|J_{\pm 1}|JM\rangle = \mp\frac{1}{2^{1/2}}[(J \mp M)(J \pm M + 1)]^{1/2}\delta_{M',M\pm 1} \qquad (4.2.17a)$$

On the other hand, using Eq. (4.2.9)

$$\langle JM'|T(J)_{1,\pm 1}|JM\rangle = (-1)^{J-M'}3^{1/2}\begin{pmatrix} J & J & 1 \\ M' & -M & \mp 1 \end{pmatrix}$$

$$= \mp\frac{3^{1/2}}{2^{1/2}}\left[\frac{(J \mp M)(J \pm M + 1)}{(2J+1)(J+1)J}\right]^{1/2}\delta_{M',M\pm 1}$$

$$(4.2.17b)$$

By composing Eqs. (4.2.16a) and (4.2.17a) with (4.2.16b) and (4.2.17b), respectively, we obtain the operator relation:

$$T(J)_{1Q} = \left[\frac{3}{(2J+1)(J+1)J}\right]^{1/2} J_Q \qquad (4.2.18)$$

Thus the vector operators $T(J)_{1Q}$ are proportional to the spherical components of the angular momentum operator.

Similarly, the second-rank tensor $T(J)_{2Q}$ can be related to quadratic combinations of the angular momentum vector components. The spherical components $T(J)_{2Q}$ of the second-rank tensor are related to the Cartesian components by the following equations (which are given without proof):

$$T(J)_{20} = \frac{N_2}{6^{1/2}} (3J_Z^2 - \mathbf{J}^2)$$

$$T(J)_{2\pm 1} = \mp \frac{N_2}{2} [(J_X J_Z + J_Z J_X) \pm i(J_Y J_Z + J_Z J_Y)] \qquad (4.2.19)$$

$$T(J)_{2\pm 2} = \frac{N_2}{2} [J_X^2 - J_Y^2 \pm i(J_X J_Y + J_Y J_X)]$$

with

$$N_2 = \left[\frac{30}{(2J + 3)(2J + 1)J(2J - 1)(J + 1)} \right]^{1/2} \qquad (4.2.20)$$

It should be noted that relations (4.2.18) and (4.2.19) hold only in the case of sharp angular momentum. The matrix elements of J_Q vanish between states $\langle J'M'|$ and $|JM\rangle$ with $J' \neq J$, whereas the elements of $T(J'J)_{KQ}$ are in general nonzero for $J' \neq J$. We will return to this point in the following section.

4.2.4. Some Important Properties of the Tensor Operators

The adjoint $T(J'J)_{KQ}^{\dagger}$ of an operator $T(J'J)_{KQ}$ is defined by expressing its matrix elements in terms of the elements (4.2.9):

$$\langle JM|T(J'J)_{KQ}^{\dagger}|J'M'\rangle = \langle J'M'|T(J'J)_{KQ}|JM\rangle^* \qquad (4.2.21)$$

In this case the star denoting the complex conjugate is superfluous because the elements (4.2.9) are real. Equation (4.2.21) defines a matrix representation of $T(J'J)_{KQ}^{\dagger}$. This is a matrix with $(2J + 1)$ rows and $(2J' + 1)$ columns which is obtained by interchanging the rows and columns of the matrix (4.2.9). In order to obtain a relation between the operators the right-hand side of Eq. (4.2.21) is transformed into a matrix with $(2J + 1)$ rows and $(2J' + 1)$ columns. Substituting Eq. (4.2.9) into Eq. (4.2.21) and using the symmetry property (C5) of the $3j$ symbol yields

$$\langle JM|T(J'J)_{KQ}^{\dagger}|J'M'\rangle = (-1)^{J'-M'} \begin{pmatrix} J' & J & K \\ M' & -M & -Q \end{pmatrix} (2K + 1)^{1/2}$$

$$= (-1)^{J'-M'} (2K + 1)^{1/2} \begin{pmatrix} J & J' & K \\ M & -M' & Q \end{pmatrix}$$

$$= (-1)^{J'-J+Q} \langle JM|T(JJ')_{K-Q}|J'M'\rangle \qquad (4.2.22)$$

where the $3j$ symbol in the second line has been expressed in terms of the elements (4.2.9). From Eq. (4.2.22) it follows that

▶ $$T(J'J)^{\dagger}_{KQ} = (-1)^{J'-J+Q}T(JJ')_{K-Q} \qquad (4.2.23)$$

where now both operators are represented by matrices with $(2J + 1)$ rows and $(2J' + 1)$ columns.

Equation (4.2.23) can be used to derive an important result. With the help of Eqs. (4.2.9) and (4.2.21) and the orthonormality condition of the $3j$ symbols we obtain

$$\text{tr } T(J'J)_{KQ}T(J'J)^{\dagger}_{K'Q'} = \sum_{M'M} \langle J'M'|T(J'J)_{KQ}|JM\rangle\langle JM|T(J'J)^{\dagger}_{K'Q'}|J'M'\rangle$$
▶
$$= \delta_{K'K}\delta_{Q'Q} \qquad (4.2.24)$$

It should be noted that the product $T(J'J)_{KQ}T(J'J)^{\dagger}_{K'Q'}$ corresponds to a square matrix with $(2J' + 1)$ rows and columns and hence the trace of the product is well defined. It follows from Eq. (4.2.24) that

$$\text{tr } T(J)_{KQ} = \text{tr } T(J)_{KQ} \cdot 1 = (2J + 1)^{1/2}\delta_{K0}\delta_{Q0} \qquad (4.2.25)$$

where we used the relation (4.2.14). *All tensors $T(J)_{KQ}$ therefore have zero trace except the monopole.*

Finally, we recall from angular momentum theory that all irreducible tensor operators V_{KQ} satisfy the Wigner–Eckart theorem:

▶ $$\langle J'M'|V_{KQ}|JM\rangle = (-1)^{J'-M'}\begin{pmatrix} J' & K & J \\ -M' & Q & +M \end{pmatrix}\langle J'\|V_K\|J\rangle \qquad (4.2.26)$$

It is important to note that the "reduced" matrix element $\langle J'\|V_K\|J\rangle$ is a scalar and independent of M', M, and Q. The $3j$ symbol, on the other hand, is a well-defined number which reflects the geometry of the interaction. The Wigner–Eckart theorem therefore separates out those quantities which depend explicitly on the dynamics of the interaction from those which are purely geometrical.

Applying Eq. (4.2.26) to the tensor operators $T(J'J)_{KQ}$ gives

$$\langle J'M'|T(J'J)_{KQ}|JM\rangle = (-1)^{J'-M'}\begin{pmatrix} J' & K & J \\ -M' & Q & M \end{pmatrix}\langle J'\|T_K\|J\rangle$$
$$(4.2.27)$$

where $\langle J'\|T_K\|J\rangle$ is the corresponding reduced matrix element. Comparing Eq. (4.2.27) with Eq. (4.2.9) and using the symmetry property of the $3j$ symbol (C5) it can be seen that

$$\langle J'\|T_K\|J\rangle = (2K + 1)^{1/2} \qquad (4.2.28)$$

Inserting Eq. (4.2.28) back into (4.2.27) shows that the tensor operators $T(J'J)_{KQ}$ are purely geometrical quantities.

4.3. State Multipoles (Statistical Tensors)

4.3.1. Definition of State Multipoles

Consider an ensemble of particles in various angular momentum states $|JM\rangle$ characterized by a density matrix ρ with elements $\langle J'M'|\rho|JM\rangle$ [see, for example, Eq. (2.3.6)]. The density operator in the $\{|JM\rangle\}$ representation can then be written in the form

$$\rho = \sum_{J'JM'M} \langle J'M'|\rho|JM\rangle|J'M'\rangle\langle JM| \qquad (4.3.1)$$

according to Eqs. (2.2.3) and (2.2.4). Substitution of Eq. (4.2.7) into Eq. (4.3.1) yields

$$\rho = \sum_{J'JKQ}\left[\sum_{M'M} \langle J'M'|\rho|JM\rangle(-1)^{J'-M'}\begin{pmatrix} J' & J & K \\ M' & -M & -Q \end{pmatrix}\right.$$
$$\left. \times (2K+1)^{1/2}\right]T(J'J)_{KQ} \qquad (4.3.2)$$

The *state multipoles* or *statistical tensors* are defined as

▶ $$\langle T(J'J)_{KQ}^\dagger\rangle = \sum_{M'M}(-1)^{J'-M'}(2K+1)^{1/2}\begin{pmatrix} J' & J & K \\ M' & -M & -Q \end{pmatrix}$$
$$\times \langle J'M'|\rho|JM\rangle \qquad (4.3.3)$$

Using Eq. (4.3.3) in Eq. (4.3.2) gives the expansion of the density operator in terms of irreducible tensor operators:

▶ $$\rho = \sum_{J'JKQ} \langle T(J'J)_{KQ}^\dagger\rangle T(J'J)_{KQ} \qquad (4.3.4)$$

Multiplying both sides of Eq. (4.3.4) by $T(J'J)_{K'Q'}^\dagger$, taking the trace, and using the relation (4.2.24) gives

▶ $$\langle T(J'J)_{KQ}^\dagger\rangle = \text{tr}\,\rho T(J'J)_{KQ}^\dagger \qquad (4.3.5)$$

which is equivalent to Eq. (4.3.3). The expression (4.3.3) can be inverted by multiplying both sides by

$$(2K+1)^{1/2}\begin{pmatrix} J' & J & K \\ N' & -N & -Q \end{pmatrix}$$

and summing over all values of K and Q. This gives

▶ $$\langle J'N'|\rho|JN\rangle = \sum_{KQ}(-1)^{J'-N'}(2K+1)^{1/2}\begin{pmatrix} J' & J & K \\ N' & -N & -Q \end{pmatrix}$$
$$\times \langle T(J'J)_{KQ}^\dagger\rangle \qquad (4.3.6)$$

The two descriptions of a system, in terms of density matrix elements and in terms of state multipoles, are therefore equivalent. They can be transformed into each other by applying Eqs. (4.3.3) and (4.3.6). Equations (4.3.3)–(4.3.6) are of great importance in all problems where angular momentum properties play a role, for example in angular correlation theory, optical pumping work, and spin-polarization phenomena. The utility of the state multipoles will become evident through the examples and discussions in the following chapters of this book.

If the ensemble of interest is an incoherent mixture of J states the density matrix is diagonal in J according to Section 2.3:

$$\langle J'M'|\rho|JM\rangle = \langle JM'|\rho|JM\rangle\delta_{J'J}$$

and from Eq. (4.3.3) follows

$$\langle T(J'J)_{KQ}^{\dagger}\rangle = \langle T(J)_{KQ}^{\dagger}\rangle\delta_{J'J}$$

Equation (4.3.4) then reduces to

$$\rho = \sum_{JKQ} \langle T(J)_{KQ}^{\dagger}\rangle T(J)_{KQ} \qquad (4.3.7)$$

This result shows that multipoles $T(J'J)_{KQ}$ with $J' \neq J$ describe the coherence between states of different angular momentum J.

If the ensemble of interest is an incoherent superposition of states with different quantum numbers M, then the density matrix is diagonal in M and Eq. (4.3.3) shows that all multipoles with $Q \neq 0$ vanish. The corresponding density operator is then given by

$$\rho = \sum_{J'JK} \langle T(J'J)_{K0}^{\dagger}\rangle T(J'J)_{K0} \qquad (4.3.8)$$

Hence the coherence between states with different quantum number M is characterized by the nonvanishing multipoles with $Q \neq 0$.

4.3.2. Basic Properties of State Multipoles

The Hermiticity condition (2.2.5) becomes

$$\langle J'M'|\rho|JM\rangle = \langle JM|\rho|J'M'\rangle^* \qquad (4.3.9)$$

for the case under discussion here. By taking the complex conjugate of Eq. (4.3.3), using Eq. (4.3.9) we obtain

▶ $$\langle T(J'J)_{KQ}^{\dagger}\rangle^* = (-1)^{J'-J+Q}\langle T(JJ')_{K-Q}^{\dagger}\rangle \qquad (4.3.10)$$

For sharp angular momentum $J' = J$ Eq. (4.3.10) implies

▶ $$\langle T(J)_{KQ}^{\dagger}\rangle^* = (-1)^{Q}\langle T(J)_{K-Q}^{\dagger}\rangle \qquad (4.3.11)$$

which relates the multipoles of components Q and $-Q$ to each other. In particular Eq. (4.3.11) ensures that the *multipoles* $\langle T(J)_{K0}^{\dagger} \rangle$ *are real numbers.*

In many cases an alternative set of parameters is used:

$$\langle T(J'J)_{KQ} \rangle = \text{tr}\, \rho T(J'J)_{KQ} \qquad (4.3.12a)$$

[employing the operator $T(J'J)_{KQ}$ instead of its adjoint]. Substitution of Eq. (4.2.23) into Eq. (4.3.12a) and use of the definition (4.3.5) yields

$$\langle T(J'J)_{KQ}^{\dagger} \rangle = (-1)^{J'-J+Q} \langle T(JJ')_{K-Q} \rangle \qquad (4.3.12b)$$

Applying Eq. (4.3.10) the two parametrizations can be seen to be related to each other by

$$\langle T(J'J)_{KQ}^{\dagger} \rangle^{*} = \langle T(J'J)_{KQ} \rangle \qquad (4.3.12c)$$

In order to see the significance of Eq. (4.3.3) we will consider the transformation properties of the state multipoles under rotations. A set of multipoles $\langle T(J'J)_{KQ}^{\dagger} \rangle$ can be defined with respect to a coordinate system with axes X, Y, Z with corresponding quantum numbers M', M, Q in Eq. (4.3.3). A second set of multipoles $\langle T(J'J)_{Kq}^{\dagger} \rangle$ can be defined with respect to the coordinate system x, y, z shown in Figure 4.1. Using Eqs. (4.3.12a) and (4.2.13),

$$\langle T(J'J)_{KQ}^{\dagger} \rangle = \langle T(J'J)_{KQ} \rangle^{*} = [\text{tr}\, \rho T(J'J)_{KQ}]^{*}$$

$$= \left[\sum_{q} D(\omega)_{qQ}^{(K)} \text{tr}\, \rho T(J'J)_{Kq} \right]^{*}$$

Using Eqs. (4.3.12a) and (4.3.12c) then gives

▶
$$\langle T(J'J)_{KQ}^{\dagger} \rangle = \sum_{q} \langle T(J'J)_{Kq}^{\dagger} \rangle D(\omega)_{qQ}^{(K)*} \qquad (4.3.13)$$

This result shows that the state multipoles transform as irreducible tensors of rank K and component Q.

4.3.3. Physical Interpretation of State Multipoles. The Orientation Vector and Alignment Tensor

The irreducible components $\langle T_{KQ}^{\dagger} \rangle$ of the density matrix in general have a deeper physical significance than the elements of ρ. In this section we will discuss the physical interpretation of the lower-rank tensors in the case of sharp angular momentum.

The tensor with rank $K = 0$ is merely a normalization constant. Taking the trace of Eq. (4.3.4) and using Eq. (4.2.25) gives

$$\langle T(J)_{00}^{\dagger} \rangle = \frac{\text{tr}\, \rho}{(2J + 1)^{1/2}} \qquad (4.3.14)$$

The three components with $K = 1$ and $Q = \pm 1, 0$ transform as the components of a vector. From Eq. (4.2.18) and the definition (4.3.5) it is readily found that

$$\langle T(J)_{1Q}^{\dagger} \rangle = \left[\frac{3}{(2J + 1)(J + 1)J} \right]^{1/2} \text{tr} \, \rho J_Q^{\dagger}$$

$$= \left[\frac{3}{(2J + 1)(J + 1)J} \right]^{1/2} \langle J_Q^{\dagger} \rangle \, \text{tr} \, \rho \qquad (4.3.15)$$

where the expectation value $\langle J_Q^{\dagger} \rangle$ of the operator $J_Q^{\dagger}$ is defined by Eq. (2.2.9a). Taking the complex conjugate of Eq. (4.3.15) and using the relation (4.3.12b) we obtain the alternative form:

$$\langle T(J)_{1Q} \rangle = \left[\frac{3}{(2J + 1)(J + 1)J} \right]^{1/2} \langle J_Q \rangle \, \text{tr} \, \rho \qquad (4.3.15a)$$

The three parameters $\langle T(J)_{1Q}^{\dagger} \rangle$ with $Q = \pm 1, 0$ are often called the components of the "orientation vector". As shown by Eq. (4.3.15) the orientation vector is proportional to the net angular momentum $\langle \mathbf{J} \rangle$ of the ensemble under discussion. Since

$$\langle \boldsymbol{\mu} \rangle = -g\mu_B \langle \mathbf{J} \rangle \qquad (4.3.16)$$

where $\langle \boldsymbol{\mu} \rangle$ denotes the magnetic dipole vector averaged over the ensemble under consideration, it can be seen that *the orientation vector is proportional to the net magnetic dipole vector of the given system* (g is the Landé factor, μ_B the Bohr magneton).

In a similar way the components $\langle T(J)_{2Q}^{\dagger} \rangle$ of the second-rank tensor can be expressed in terms of quadratic combinations of the angular momentum components using Eqs. (4.2.19) and (4.3.5). For example,

$$\langle T(J)_{20}^{\dagger} \rangle = \frac{N_2}{6^{1/2}} \langle 3J_Z^2 - \mathbf{J}^2 \rangle \, \text{tr} \, \rho \qquad (4.3.17)$$

The tensor $\langle T(J)_{20}^{\dagger} \rangle$ is called the *alignment tensor*. Its physical significance follows from the fact *that the components $\langle T(J)_{2Q}^{\dagger} \rangle$ are proportional to the spherical components $\langle Q_{2Q} \rangle$ of the electric quadrupole tensor*. This can be seen in the following way. The expectation value $\langle Q_{2Q} \rangle$ is defined by Eq. (2.2.9a). Applying the Wigner–Eckart theorem to the matrix elements of Q_{2Q} gives

$$\langle Q_{2Q} \rangle \, \text{tr} \, \rho = \text{tr} \, \rho Q_{2Q}$$

$$= \sum_{M'M} \langle JM' | \rho | JM \rangle \langle JM | Q_{2Q} | JM' \rangle$$

$$= \langle J \| Q \| J \rangle \sum_{M'M} \langle JM' | \rho | JM \rangle (-1)^{J-M} \begin{pmatrix} J & 2 & J \\ -M & Q & M' \end{pmatrix}$$

Using the symmetry properties of the $3j$ symbols and substitution of Eq. (4.3.3) then gives

$$\langle Q_{2Q}\rangle \operatorname{tr}\rho = \frac{\langle J\|Q\|J\rangle}{5^{1/2}}\langle T(J)_{2Q}\rangle \tag{4.3.18}$$

where the reduced matrix element is proportional to the quadrupole moment of the system (see, for example, Edmonds, 1957).

Equation (4.3.18) illustrates how the expansion of ρ in terms of irreducible components, combined with the Wigner–Eckart theorem, enables the geometrical and dynamical properties of the system to be separated. The reduced matrix element contains all information on the dynamics while the tensor $\langle T(J)_{2Q}\rangle$ describes the geometrical properties of the relevant ensemble. This aspect of the theory will be of increasing importance to our subsequent discussions.

Finally, we give the following definitions.

A system is

▶ *oriented* if at least one of the components of the orientation vector is nonvanishing, *aligned* if at least one of the components of the alignment tensor is nonzero, or *polarized* if at least one of the multipoles with $K \neq 0$ is nonzero.

The results given here hold only for sharp J. An interpretation of multipoles with $J' \neq J$ will be given in Section 4.6.

4.4. Examples: Spin Tensors

4.4.1. Spin Tensors for Spin-1/2 Particles

We start by reexamining the description of spin-1/2 particles characterized by a density matrix ρ with elements $\langle m'|\rho|m\rangle$. We define a set of state multipoles $\langle T(S)^\dagger_{KQ}\rangle$, the so-called *spin tensors*, by means of Eq. (4.3.3). Thus, for $S = 1/2$, we write

$$\langle T(1/2)^\dagger_{KQ}\rangle = \sum_{m'm} (-1)^{(1/2)-m'}(2K+1)^{1/2}\begin{pmatrix} 1/2 & 1/2 & K \\ m' & -m & -Q \end{pmatrix}\langle m'|\rho|m\rangle \tag{4.4.1}$$

Because of condition (4.2.4) only terms with $K = 0$ and $K = 1$ are allowed in Eq. (4.4.1). The monopole with $K = 0$ is a normalization constant. If the spin-1/2 density matrix is normalized so that $\operatorname{tr}\rho = 1$ as in Section 1.1, then using Eq. (4.3.14) the monopole term is given by

$$\langle T(1/2)_{00}\rangle = 1/2^{1/2} \tag{4.4.2}$$

The three vector components $\langle T(1/2)^\dagger_{1Q}\rangle$ are related to the corresponding components of the spin vector by Eq. (4.3.15a):

$$\langle T(1/2)^\dagger_{1Q}\rangle = 2^{1/2}\langle S_Q\rangle \tag{4.4.3a}$$

where S_Q denotes the Qth spherical component of the spin operator S as defined by Eq. (4.2.15). Using the definition of the Pauli matrices $(1/2)\sigma_i = S_i$ $(i = x, y, z)$ and of the polarization vector $\mathbf{P}$ we obtain

$$\langle T(1/2)^\dagger_{1Q}\rangle = P_Q/2^{1/2} \tag{4.4.3b}$$

Thus *the state multipoles (spin tensors)* $\langle T(1/2)^\dagger_{1Q}\rangle$ *are proportional to the spherical components of the polarization vector* defined by Eq. (4.2.15):

$$P_{\pm 1} = \mp(1/2^{1/2})(P_x \pm iP_y), \qquad P_0 = P_z \tag{4.4.4}$$

The expansion of the spin-1/2 density matrix in terms of spin tensors can be obtained by applying Eq. (4.3.4):

$$\rho = \sum_{KQ} \langle T(1/2)^\dagger_{KQ}\rangle T(1/2)_{KQ}$$

$$= (1/2)\mathbf{1} + \sum_Q \langle T(1/2)^\dagger_{1Q}\rangle T(1/2)_{1Q} \tag{4.4.5}$$

which is simply a reformulation of Eq. (1.1.45).

4.4.2. Description of Spin-1 Particles

Spin-1 particles are described by three basic states corresponding to the three possible eigenvalues of the operator S_z. These states can be represented in form of the three-dimensional column vectors

$$\begin{pmatrix} 1 \\ 0 \\ 0 \end{pmatrix}, \quad \begin{pmatrix} 0 \\ 1 \\ 0 \end{pmatrix}, \quad \begin{pmatrix} 0 \\ 0 \\ 1 \end{pmatrix} \tag{4.4.6a}$$

In the standard representation (4.4.6a) the operators S_x, S_y, S_z are given by the matrices

$$S_x = \frac{1}{2^{1/2}}\begin{pmatrix} 0 & 1 & 0 \\ 1 & 0 & 1 \\ 0 & 1 & 0 \end{pmatrix}, \quad S_y = \frac{i}{2^{1/2}}\begin{pmatrix} 0 & -1 & 0 \\ 1 & 0 & -1 \\ 0 & 1 & 0 \end{pmatrix}, \quad S_z = \begin{pmatrix} 1 & 0 & 0 \\ 0 & 0 & 0 \\ 0 & 0 & -1 \end{pmatrix}$$

$$\tag{4.4.6b}$$

The components of the polarization vector of spin-S particles are defined by the relation

$$P_i = \frac{\langle S_i \rangle}{S} \tag{4.4.7}$$

($i = x, y, z$). For $S = 1/2$ Eq. (4.4.7) reduces to Eq. (1.1.4). For $S = 1$ we have

$$P_i = \langle S_i \rangle \tag{4.4.7a}$$

It is instructive to calculate $\mathbf{P}$ for the three basis states $|+1\rangle$, $|0\rangle$, $|-1\rangle$. Using the explicit representations (4.4.6a) and (4.4.6) and performing the calculations in the same way as in Section 1.1.2 it is found that $P_x = P_y = 0$ in all three cases and $P_z = 1$, $P_z = 0$, $P_z = -1$ for the states $|+1\rangle$, $|0\rangle$, $|-1\rangle$, respectively.

It is important to note that the state $|0\rangle$ has a polarization vector of zero magnitude. This shows a fundamental difference between spin-1/2 and spin-1 particles: For $S = 1$ it is possible to have a pure spin state which does not have a preferred direction (that is, there need not be a direction in which the spins are "pointing"). This is easily understood in terms of the semiclassical vector model. The state $M = 0$ is represented by a spin vector perpendicular to the z axis and precessing around it. Clearly, in this case there can be a preferred *axis* (z axis) but it is not possible to specify a *direction* along this axis. *It is this property in particular which makes it necessary to consider quantities of higher rank than the polarization vector.* The quantities which are required must not depend on the direction of the z axis, and hence quadratic combinations like $\langle S_z^2 \rangle$ or, more generally, components of second-rank tensors can be used.

Furthermore, if a beam of particles in the pure state $|0\rangle$ is compared with a mixture of $N_+ = N/2$ particles in the state $|+1\rangle$ and $N_- = N/2$ particles in the state $|-1\rangle$, then in both cases $\mathbf{P} = 0$. Thus knowledge of the polarization vector alone is insufficient for a complete specification of spin-1 particles and additional parameters must be introduced.

The most systematic way to obtain all the necessary parameters is to construct the relevant spin tensors $\langle T(S)_{KQ}^{\dagger} \rangle$ for $S = 1$. If the elements of the spin-1 density matrix are denoted by $\langle M'|\rho|M \rangle$, then from Eq. (4.3.3)

$$\langle T(1)_{KQ}^{\dagger} \rangle = \sum_{M'M} (-1)^{1-M'} (2K+1)^{1/2} \begin{pmatrix} 1 & 1 & K \\ M' & -M & -Q \end{pmatrix} \langle M'|\rho|M \rangle \tag{4.4.8}$$

Because of condition (4.2.4) it is necessary to construct a monopole, a vector, and a second-rank tensor. The monopole is specified by the normalization:

$$\langle T(1)^\dagger_{00} \rangle = 1/3^{1/2} \qquad (4.4.9)$$

The corresponding density matrix can then be written in the form

$$\rho = (1/3)\mathbf{1} + \sum_Q \langle T(1)^\dagger_{1Q} \rangle T(1)_{1Q} + \sum_Q \langle T(1)^\dagger_{2Q} \rangle T(1)_{2Q} \qquad (4.4.10)$$

Consequently, if the Hermiticity condition (4.3.11) is taken into account, the most general spin-1 density matrix is completely specified in terms of eight real parameters (nine if the normalization $\operatorname{tr}\rho = 1$ is dropped and $\operatorname{tr}\rho$ considered as a parameter to be determined experimentally).

As an example consider an incoherent mixture of N_+ particles in the state $|+1\rangle$, N_- particles in the state $|-1\rangle$, and N_0 particles in the state $|0\rangle$. In this case the density matrix is diagonal in the representation (4.4.5):

$$\langle M'|\rho|M \rangle = W_M \delta_{M'M} \qquad (4.4.11)$$

where $W_M = N_M/N$ and N is the total number of particles. Substitution of Eq. (4.4.11) into Eq. (4.4.8) yields

$$\langle T(1)^\dagger_{20} \rangle = (1/12^{1/2})(W_{+1} + W_{-1} - 2W_0) = \frac{N_{+1} + N_{-1} - 2N_0}{N(12)^{1/2}} \qquad (4.4.12)$$

for the tensor polarization and

$$\langle T(1)^\dagger_{10} \rangle = \frac{W_{+1} - W_{-1}}{2^{1/2}} = \frac{N_+ - N_-}{2^{1/2}N} \qquad (4.4.13)$$

for the vector polarization. (All components with $Q \neq 0$ are zero.)

The components $\langle T(S)^\dagger_{1Q} \rangle$ are proportional to the spherical components of the polarization vector. Substituting $\mathbf{S}$ for $\mathbf{J}$ into Eq. (4.3.15a) and taking definition (4.4.7a) into account gives

$$\langle T(1)^\dagger_{1Q} \rangle = (1/2^{1/2})P_Q \qquad (4.4.14)$$

where the spherical components of $\mathbf{P}$ are defined by Eq. (4.4.4). The five components of the second-rank tensor $\langle T(S)^\dagger_{2Q} \rangle$ can be constructed from quadratic combinations of the spin operators S as in Eq. (4.2.19) with $J_i = S_i$ and $N_2 = 1$.

The use of Cartesian tensors in actual calculations can have some advantages but the spherical tensors $\langle T(S)_{KQ}^{\dagger} \rangle$ have a simpler algebra, which in general considerably simplifies the calculations.

Finally, let us consider the consequences of condition (2.2.12). Substitution of Eq. (4.4.10) for ρ and use of Eq. (4.2.24) yields $\text{tr}\,\rho = 1$ and

$$\text{tr}\,(\rho^2) = \sum_{\substack{KQ \\ K'Q'}} \langle T(1)_{K'Q'}^{\dagger} \rangle \langle T(1)_{KQ}^{\dagger} \rangle \, \text{tr}\,[T(1)_{K'Q'} T(1)_{KQ}]$$

$$= \sum_{KQ} (-1)^Q \langle T(1)_{KQ}^{\dagger} \rangle \langle T(1)_{K-Q}^{\dagger} \rangle$$

$$= \sum_{KQ} |\langle T(1)_{KQ}^{\dagger} \rangle|^2$$

where Eqs. (4.2.23), (4.2.24), and (4.3.11) have been used. Thus the spin tensors are restricted such that

$$\sum_{KQ} |\langle T(1)_{KQ}^{\dagger} \rangle|^2 \leq 1 \qquad (4.4.15)$$

If (and only if) ρ characterizes a pure state then

$$\sum_{KQ} |\langle T(1)_{KQ}^{\dagger} \rangle|^2 = 1 \qquad (4.4.16)$$

In the literature a beam of spin-1 particles is usually referred to as *completely polarized* if the beam is in a pure spin state, that is, if (and only if) condition (4.4.16) is satisfied. In this case the state of the beam can be represented by a single-state vector $|\chi\rangle$ which can be expanded in terms of the basic states (4.4.5):

$$|\chi\rangle = a_{+1}|+1\rangle + a_0|0\rangle + a_{-1}|-1\rangle \qquad (4.4.17)$$

Equation (4.4.17) shows that in general a completely polarized beam of spin-1 particles is specified by five real parameters, for example the magnitudes of the coefficients, a_M, and their relative phases.

The formalism which has been developed here is of considerable interest in the description of scattering processes involving polarized particles. This topic will not be discussed in this book except for a few relevant formulas which are given in Appendixes A and B. Detailed accounts can be found in textbooks on scattering theory, for example in the books by Rodberg and Thaler (1967) or Burke and Joachain (est. 1982). Discussion of scattering experiments using polarized electrons, including many experimental details, can be found in Kessler (1976). The more formal aspects of the theory have been discussed by Robson (1974) (see also Blum and Kleinpoppen, to be published).

4.5. Symmetry Properties. Relation between Symmetry and Coherence

4.5.1. Axially Symmetric Systems

4.5.1.1. General Results

The excitation of an ensemble of particles (atoms or nuclei) can be achieved in several ways: by the interaction of external fields, by absorption of radiation, by impact by other particles, and so on. Let us assume that the excitation process is axially symmetric with respect to some axis. This axis can be, for example, the direction of an external field. In excitation by electron impact, in which the scattered electrons are not detected, the symmetry axis would be defined by the direction of the initial electron beam.

Throughout this section we will always take the symmetry axis as Z axis of our coordinate system (quantization axis). The choice of the X and Y axes perpendicular to Z is arbitrary and *the physical properties of the ensemble must therefore be independent of this choice* (the particles cannot "know" how the X and Y axes are defined). In particular, the real and imaginary parts of the state multipoles are directly measurable quantities (see Chapter 5) and hence must have the *same numerical values* in the XYZ system and in any xyZ system which can be obtained by a rotation about the Z axis through an arbitrary angle γ. This gives the following *symmetry condition*:

$$\langle T(J'J)^{\dagger}_{KQ}\rangle = \langle T(J'J)^{\dagger}_{KQ}\rangle_{\text{rot}} \tag{4.5.1}$$

where $\langle T(J'J)^{\dagger}_{KQ}\rangle$ and $\langle T(J'J)^{\dagger}_{KQ}\rangle_{\text{rot}}$ are defined in the fixed XYZ and the rotated xyZ systems, respectively. Equation (4.5.1) relates two complex quantities and hence real and imaginary parts are equal.

On the other hand, from the transformation law (4.3.13), the multipoles are related by

$$\langle T(J'J)^{\dagger}_{KQ}\rangle = \sum_q \langle T(J'J)^{\prime}_{Kq}\rangle_{\text{rot}} \cdot D(00\gamma)^{(K)*}_{qQ} \tag{4.5.2}$$

where γ denotes the angle between the X and the x axes. The elements of the rotation matrix specifying the rotation around Z are given by

$$D(00\gamma)^{(K)*}_{qQ} = \exp\left(-iQ\gamma\right)\delta_{qQ} \tag{4.5.3}$$

[see Eq. (C12)]. Substitution of Eq. (4.5.3) into (4.5.2) yields

$$\langle T(J'J)^{\dagger}_{KQ}\rangle = \langle T(J'J)^{\dagger}_{KQ}\rangle_{\text{rot}} \exp\left(-iQ\gamma\right) \tag{4.5.4}$$

Equation (4.5.4) is the general transformation law which holds for *any* angle γ. In addition, because of the axial symmetry, the condition (4.5.1) must also be satisfied for *any* angle γ. This is only possible if $Q = 0$. Thus *axially symmetric systems are characterized by multipole components* $\langle T(J'J)_{K\,0}^{\dagger}\rangle$, all components with $Q \neq 0$ are necessarily zero because they violate the symmetry condition (4.5.1). The density operator characterizing systems with axial symmetry is therefore given by

$$\rho = \sum_{J'JK} \langle T(J'J)_{K\,0}^{\dagger}\rangle T(J'J)_{K\,0} \qquad (4.5.5)$$

It follows from Eq. (4.3.8) that ρ is diagonal in M. We therefore have the following general result. If an ensemble of particles has been excited by a process, which is axially symmetric with respect to a preferred axis, *then states with different components of angular momentum are necessarily incoherently excited* (provided the quantization axis coincides with the symmetry axis). *The production of a coherent superposition of states with different angular momentum components requires an excitation process which is not axially symmetric.*

4.5.1.2. Reversal of the Z Axis

In this and the following section we will consider states with sharp angular momentum $J' = J$.

Axially symmetric systems can be classified by their transformation properties under a reversal of the symmetry axis: $Z \rightarrow (-Z)$. This corresponds to a rotation around the Y axis by an angle π. The corresponding rotation matrix elements are given by Eqs. (C12) and (C15):

$$D(0\pi 0)_{qQ}^{(K)} = (-1)^{K+Q}\delta_{q-Q} \qquad (4.5.6)$$

Inserting Eq. (4.5.6) into the general relation (4.3.14) and accounting for the axial symmetry we obtain

$$\langle T(J)_{K\,0}^{\dagger}\rangle = \sum_{q} \langle T(J)_{Kq}^{\dagger}\rangle_{\text{rot}}(-1)^{K+q}\delta_{q0}$$

$$= (-1)^{K}\langle T(J)_{K\,0}^{\dagger}\rangle_{\text{rot}} \qquad (4.5.7)$$

where $\langle T(J)_{K\,0}^{\dagger}\rangle$ and $\langle T(J)_{K\,0}^{\dagger}\rangle_{\text{rot}}$ are defined with respect to the Z and $-Z$ axes.

If a given ensemble is invariant with respect to the transformation $Z \rightarrow -Z$, that is, the values of all measurable quantities remain unchanged under this operation, then this requires that

$$\langle T(J)_{K\,0}^{\dagger}\rangle = \langle T(J)_{K\,0}^{\dagger}\rangle_{\text{rot}} \qquad (4.5.8)$$

The expressions (4.5.7) and (4.5.8) can only be simultaneously satisfied by multipoles of even rank K. As a result an axially symmetric system which is

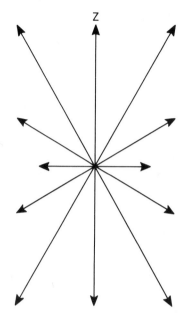

Figure 4.2. Aligned axially symmetric system.

invariant under the reversal of the symmetry axis *is characterized by multipoles of even rank K* and all tensors with K odd are necessarily zero. In particular the orientation vector vanishes. It follows that the system under consideration is a special case of an *aligned system* as defined in Section 4.3.3.

Let us consider the effect of the symmetry condition (4.5.8) on the density matrix elements. Applying Eq. (4.3.6) and the symmetry properties of the $3j$ symbols we have

$$\langle J - M |\rho| J - M \rangle = \sum_K \langle T(J)^{\dagger}_{K\,0}\rangle (2K + 1)^{1/2}(-1)^{J+M}\begin{pmatrix} J & J & K \\ -M & M & 0 \end{pmatrix}$$

$$= \langle JM |\rho| JM \rangle \qquad\qquad (4.5.9)$$

since only multipoles with even K contribute. The diagonal elements of ρ are proportional to the populations of the state $|JM\rangle$ and the proportionality constant is specified by the normalization. Equation (4.5.9) shows that the states $|JM\rangle$ and $|J - M\rangle$ are *equally populated*.

It will be useful to discuss these results from a more physical viewpoint. Semiclassically, with a state $|JM\rangle$ is associated a vector of length $[J(J + 1)]^{1/2}$ with its Z component M precessing around Z. The length of the vectors can be changed without altering their directions in space, in such a way that the length is made proportional to the number of particles in the corresponding state $|JM\rangle$. In terms of this model a system satisfying Eq. (4.5.9) can be represented diagrammatically by Figure 4.2 where the arrows represent the angular momentum vectors, pointing into those directions of space which are allowed. The figure is axially symmetric and invariant under the operation $Z \to -Z$, that is, vectors pointing in opposite directions have the same length. In particular, the figure shows that the net angular momentum $\langle \mathbf{J} \rangle$ of an aligned system is zero.

The condition (4.5.9) is trivially fulfilled for an atomic ensemble where all the particles are in the same state $|J, M = 0\rangle$. A system of this kind is a particularly simple example of an aligned system without orientation.

4.5.1.3. Oriented Systems

An axially symmetric system being not invariant against a reversal of the symmetry (Z) axis is illustrated in Figure 4.3. In this case, the length of vectors pointing in opposite directions is different. The figure shows that the system possesses a nonvanishing net angular momentum component. The excess of angular momentum vectors pointing in one direction can be described either by $\langle J_Z \rangle$ or $\langle T(J)_{10} \rangle$. This is an example of an oriented system.

Oriented systems can be produced, for example, by excitation of atoms or nuclei by circularly polarized light which populates the states $|JM\rangle$ and $|J, -M\rangle$ differently because of the angular momentum selection rules.

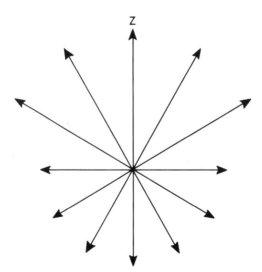

Figure 4.3. Oriented axially symmetric system.

4.5.2. Spherically Symmetric Systems

Consider an ensemble of particles which has no preferred axis in space (for example, a system of unpolarized spin-1/2 particles). In this case the XYZ axes of our coordinate system can be chosen arbitrarily and consequently all physical properties of the ensemble must be independent of the positions of the coordinate axes. This requirement gives the symmetry condition

$$\langle T(J'J)_{KQ}^{\dagger} \rangle = \langle T(J'J)_{KQ}^{\dagger} \rangle_{\text{rot}} \tag{4.5.10}$$

for *all* rotated systems, that is, for *any* choice of the Euler angles. From an inspection of Eq. (4.3.14) it can be seen that the condition (4.5.10) can only be satisfied if $D(\omega)_{qQ}^{(K)}$ is proportional to the unit matrix, that is, if all multipoles except the monopole with $K = 0$ vanish. Spherically symmetric systems are therefore characterized by the monopole $\langle T(J'J)_{00}^{\dagger} \rangle$ only. From the angular momentum coupling rules (4.2.4) it follows that a tensor with $K = 0$ can only be constructed if $J' = J$, and hence from Eq. (4.3.7) it follows that the corresponding density matrix is diagonal in J. In particular, if all particles of the given ensemble have the same momentum J the density matrix is given by

$$\rho = \langle T(J)_{00} \rangle T(J)_{00} = \frac{\text{tr}\,\rho}{2J + 1}\,\mathbf{1} \tag{4.5.11}$$

where we used Eqs. (4.2.14) and (4.3.14). In conclusion it has been shown that spherically symmetric systems *are necessarily incoherent in J and M*. For a given J the magnetic substates are equally populated. In the semiclassical model spherically symmetric systems can be represented by an isotropic distribution of angular momentum vectors as shown in Figure 4.4.

4.5.3. Examples: Photoabsorption by Atoms (Nuclei)

We will now illustrate the theory described in the preceding sections with some examples. Consider an ensemble of atoms or nuclei which is initially in a state with angular momentum $J_0 = 0$, and is then excited by photon absorption to a state with $J = 1$. We will first discuss the case in which the incident light is *unpolarized*. The total system is axially symmetric with respect to the direction of propagation $\mathbf{n}$ of the light, and it is therefore convenient to choose this as the quantization axis. Consequently, the excited state density matrix is diagonal: $\langle M'|\rho|M \rangle = \langle M|\rho|M \rangle \delta_{M'M}$ with respect to $\mathbf{n}$. The incident light beam can be considered as an incoherent mixture of the two helicity states each with equal intensity, and hence only atomic states with $M = \pm 1$ can be excited and $\langle 0|\rho|0 \rangle = 0$. Furthermore, as a consequence

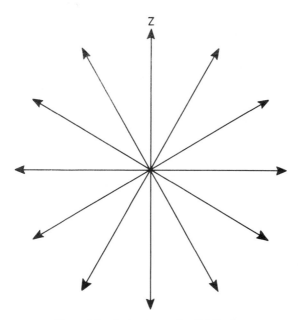

Figure 4.4. Isotropic angular distribution.

of angular momentum conservation and the fact that the light beam has equal components of the two helicity states, the atomic states $|+1\rangle$ and $|-1\rangle$ are equally populated, i.e., $\langle +1|\rho|+1\rangle = \langle -1|\rho|-1\rangle$, and the net angular momentum component of the atomic states is zero. Thus no net angular momentum is transferred to the atoms and the state is aligned but not oriented. In terms of state multipoles, the atomic state can be completely specified by two parameters only, the monopole $\langle T_{00}\rangle$ and the alignment parameter $\langle T_{20}^{\dagger}\rangle$.

Consider now the case that the incident radiation is *linearly polarized*. The Z axis can be chosen in such a way that Z is parallel to the electric vector **E**. Absorption of this light will occur through π transitions ($\Delta M = 0$) which will produce an alignment but no orientation in the excited atoms. Thus *with the quantization axis parallel to E* the excited ensemble is again characterized in terms of two parameters, $\langle T_{00}\rangle$ and $\langle T_{20}^{\dagger}\rangle$.

Finally, if the incident light is *circularly polarized* and the direction of motion is chosen as the quantization axis, then the states $|+1\rangle$ and $|-1\rangle$ will not be equally populated. The light will produce an orientation and the atomic density matrix is specified by three parameters, $\langle T_{00}\rangle$, $\langle T_{10}^{\dagger}\rangle$, $\langle T_{20}^{\dagger}\rangle$, where $\langle T_{10}^{\dagger}\rangle$ gives the net amount of angular momentum which is transferred to the atoms.

4.6. Excitation of Atoms by Electron Impact II. State Multipoles

4.6.1. Collisional Production of Atomic Orientation

In Section 3.5 an expression was derived for the density matrix ρ describing an atomic ensemble excited by electrons which were "scattered" in a fixed direction (with momentum $\mathbf{p}_1$). We will now give a description of the excited atoms in terms of state multipoles. This will be a convenient starting point for the discussions in Chapter 6.

The elements of the density matrix, averaged over all spins, are given by Eq. (3.5.17). Using Eq. (4.3.3) we can define multipole components $\langle T(L)_{KQ}^\dagger \rangle$ which describe the orbital states alone:

$$
\langle T(L)_{KQ}^\dagger \rangle = \sum_{M'M} (-1)^{L-M'}(2K+1)^{1/2}\begin{pmatrix} L & L & K \\ M' & -M & -Q \end{pmatrix}\langle LM'|\rho|LM\rangle
$$

$$
= \sum_{M'M} (-1)^{L-M'}(2K+1)^{1/2}\begin{pmatrix} L & L & K \\ M' & -M & -Q \end{pmatrix}\langle f_M, f_M^* \rangle
$$

$$(4.6.1)$$

In accordance with condition (4.2.4), for a given orbital angular momentum L all tensors of rank $K = 0, 1, \ldots, 2L+1$ and components $|Q| \le K$ must be constructed in order to obtain a complete description of the atomic ensemble.

As discussed in Section 3.5.2 the atomic system under consideration must be invariant under reflections in the scattering $(X-Z)$ plane. In order to see the consequences of this symmetry requirement on the state multipoles we will first consider the orientation vector which is proportional to the net angular momentum $\langle \mathbf{L} \rangle$ of the atomic subensemble. Since initially the atoms were assumed to be unoriented $\langle \mathbf{L} \rangle$ is the net angular momentum transferred to the atoms during the scattering. We will now discuss the transformation properties of $\mathbf{L}$.

Suppose the orientation vector has a nonvanishing component $\sim \langle L_X \rangle$ in the X direction as shown in Figure 4.5a. A reflection in the $X-Z$ plane can be generated by rotating the system around the Y axes by an angle π followed by an inversion through the origin. Angular momentum vectors transform as *axial* vectors. Polar vectors, such as momentum, and axial vectors have the same transformation properties under rotating but behave differently under inversion: polar vectors change their sign whereas axial vectors do not. Under reflection in the scattering plane Figure 4.5a is therefore transformed into Figure 4.5b. Since the atomic ensemble must be invariant under reflection in the scattering plane the situations depicted in

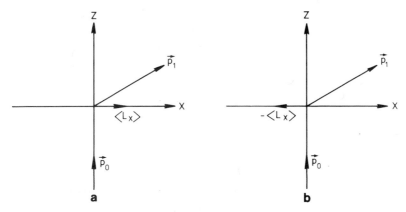

Figure 4.5. Reflection of $\langle \mathbf{L} \rangle$ in the scattering plane.

Figures 4.5a and 4.5b must occur with equal probability. The net component $\langle L_X \rangle$ must therefore vanish. The same argument holds for the component $\langle L_Z \rangle$. In this case only $\langle L_Y \rangle$, which is perpendicular to the scattering plane, can be nonzero.

In general the excited atomic system can therefore be expected to be oriented. A suggestive classical model of the mechanism responsible for the atomic orientation is that of a grazing impact illustrated in Figure 4.6. The

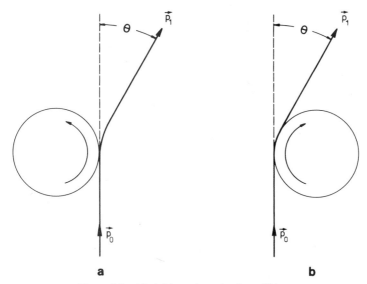

Figure 4.6. Model for orientation by collisions.

diagrams suggest that the atoms obtain an angular momentum perpendicular to the scattering plane which is opposite for a repulsive force (Figure 4.6a) and for an attractive one (Figure 4.6b).

This relation between the sign of the orientation, the deflection of the scattered particles, and the effective interactions was further studied by Fano and Komoto (1977) and Herman and Hertel (1979). They succeeded in showing that the orientation is reversed by a sign reversal of the interaction.

These results enable some conclusions to be made on the behavior of the orientation vector with respect to the scattering angle θ at fixed energy. Consider, for example, the excitation of ^1P states of helium. Electrons scattered in the forward or backward direction cannot transfer a net angular momentum to the atoms and $\langle \mathbf{L} \rangle = 0$ for $\theta° = 0°$ and $\theta = 180°$. Scattering into small angles is dominated by the long-range attractive force due to the atomic polarizability, while large-angle scattering is dominated by the short-range repulsive force caused by the atomic electrons. $\langle \mathbf{L} \rangle$ can therefore be expected to have opposite signs at small and large angles and to vanish at an intermediate angle where the contributions from the attractive and repulsive forces are equal in magnitude. These conclusions are confirmed by the recent measurements of Hollywood *et al.* (1979) and Steph and Golden (1980).

4.6.2. General Consequences of Reflection Invariance

We will now discuss the transformation properties of the state multipoles (4.6.1) under reflection in the X–Z plane. Reflection invariance of the atomic system under consideration implies that the elements of ρ must satisfy condition (3.5.11):

$$\langle LM'|\rho(L)|LM \rangle = (-1)^{M'+M}\langle L-M'|\rho(L)|L-M \rangle \qquad (4.6.2)$$

In particular, for the diagonal elements we have

$$\sigma(M) = \sigma(-M) \qquad (4.6.2a)$$

Substituting condition (4.6.2) into Eq. (4.6.1) gives

$$\langle T(L)^\dagger_{KQ} \rangle = \sum_{M'M} (-1)^{L-M'}(2K+1)^{1/2} \begin{pmatrix} L & L & K \\ M' & -M & -Q \end{pmatrix}$$
$$\times (-1)^{M'+M}\langle L-M'|\rho|L-M \rangle \qquad (4.6.3a)$$

Since all the values of M' and M are summed over in Eq. (4.6.3a) we can replace $(+M')$ and $(+M)$ by $(-M')$ and $(-M)$, respectively. Applying the symmetry property (C5c) (from Appendix C), then gives

$$\langle T(L)_{KO}^{\dagger}\rangle = (-1)^{K+Q} \sum_{M'M} (-1)^{L-M'}(2K+1)^{1/2}\begin{pmatrix} L & L & K \\ M' & -M & Q \end{pmatrix}$$

$$\times \langle LM'|\rho(L)|LM\rangle$$

$$= (-1)^{K+Q}\langle T(L)_{K-Q}^{\dagger}\rangle \tag{4.6.3}$$

From this relation and the Hermiticity condition (4.3.11) it follows that

$$\langle T(L)_{KO}^{\dagger}\rangle = (-1)^{K}\langle T(L)_{KO}^{\dagger}\rangle^{*} \tag{4.6.4}$$

In the case of the atomic system under consideration reflection in the X–Z plane has the following consequences on the state multipoles: *for even K the tensors $\langle T(L)_{KO}^{\dagger}\rangle$ are real and for K odd the tensors $\langle T(L)_{KO}^{\dagger}\rangle$ are imaginary. The components with $Q = 0$ vanish if K is odd.*

The components of orientation vector and alignment tensor are related to the corresponding angular momentum tensors by Eqs. (4.3.15) and (4.3.18). Because of the symmetry condition (4.6.4) the real or imaginary part of these expressions is zero depending on whether K is odd or even. Applying the normalization (3.5.8)

$$\text{tr}\,\rho = \sigma$$

we have for the components of the orientation vector

$$-i\langle T(L)_{1\pm1}^{\dagger}\rangle = \frac{\mp 3^{1/2}}{[2(2L+1)(L+1)L]^{1/2}}\,\sigma\langle L_Y\rangle \tag{4.6.5a}$$

$$\langle T(L)_{10}^{\dagger}\rangle = 0 \tag{4.6.5b}$$

and for the components of the alignment tensor

$$\langle T(L)_{20}^{\dagger}\rangle = \frac{N_2}{6^{1/2}}\,\sigma\langle 3L_Z^2 - \mathbf{L}^2\rangle$$

$$\langle T(L)_{2\pm1}^{\dagger}\rangle = \mp\frac{N_2}{2}\,\sigma\langle L_X L_Z + L_Z L_X\rangle \tag{4.6.6}$$

$$\langle T(L)_{2\pm2}^{\dagger}\rangle = \frac{N_2}{2}\,\sigma\langle L_X^2 - L_Y^2\rangle$$

It should be noted that the Hermiticity condition (4.3.11) restricts the number of independent multipoles. The orientation is specified in terms of one parameter [for example, $\langle T(L)_{1,+1}^{\dagger}\rangle$] and the alignment is completely characterized in terms of three independent parameters [for example, the components $\langle T(L)_{2O}^{\dagger}\rangle$ with $Q = 0, +1, +2$].

The parameters (4.6.5) and (4.6.6) are closely related to a set of quantities introduced by Fano and Macek (1973):

$$0_{1^-} = \frac{\langle L_Y \rangle}{L(L+1)}, \qquad A_{1^+} = \frac{\langle L_X L_Z + L_Z L_X \rangle}{L(L+1)}$$

$$A_0 = \frac{\langle 3L_Z^2 - L^2 \rangle}{L(L+1)}, \qquad A_{2^+} = \frac{\langle L_X^2 - L_Y^2 \rangle}{L(L+1)} \qquad (4.6.7)$$

where 0_{1^-} characterizes the orientation and the three other ones the alignment. Note, however, that the use of the relations (4.5.5), (4.6.6), and (4.6.7) is only meaningful if atomic states with sharp angular momentum L have been excited.

If atomic states with different L have been coherently excited then it is necessary to construct state multipoles $\langle T(L'L)_{KQ}^\dagger \rangle$ in order to describe completely the atomic ensemble. The transformation properties of the tensors $\langle T(L'L)_{KQ}^\dagger \rangle$ under reflection and inversion depend on whether $L' + L$ is even or odd. For example, if $L' + L$ is odd the vector $\langle T(L'L)_{1Q}^\dagger \rangle$ transforms as a polar vector and must therefore lie in the scattering plane in order to be invariant under reflection in this plane. By applying the Wigner–Eckart theorem it can be shown that this vector is related to the components $\langle r_Q \rangle$ of the net electric dipole vector induced in the atomic ensemble. We will not go further into the details of this analysis here but a more complete treatment can be found in Section 4.4 of the review by Blum and Klein-poppen (1979).

4.6.3. Axially Symmetric Atomic Systems

We will now specialize the results of the preceding sections to the case in which the scattered electrons are not observed. In this case a single axis (the direction of $\mathbf{p}_0$) is defined by the geometry of the experiment. Consequently, the excited atomic ensemble must be invariant with respect to rotations around this axis and the results of Section 4.5.1.1 apply: All multipoles with $Q \neq 0$ vanish. Denoting the corresponding density matrix by ρ then

$$\langle LM | \rho | LM \rangle = Q(M) \qquad (4.6.8)$$

where $Q(M)$ is the total cross section for excitation of the substate with magnetic quantum number M. From relation (4.6.2a) it follows that

$$Q(M) = Q(-M) \qquad (4.6.9)$$

The monopole is given by

$$\langle T(L)_{00} \rangle = \frac{Q}{(2L+1)^{1/2}} \qquad (4.6.10)$$

where $Q = \sum_M Q(M)$ is the total cross section. Substituting Eq. (4.6.8) into Eq. (4.6.1) and using (4.6.9) gives

$$\langle T(L)_{10}^{\dagger} \rangle = 0 \qquad (4.6.11a)$$

which is a consequence of (4.6.9) and

$$\langle T(L)_{20}^{\dagger} \rangle = 5^{1/2} \sum_M (-1)^{L-M} \begin{pmatrix} L & L & 2 \\ M & -M & 0 \end{pmatrix} Q(M)$$

$$= \frac{5^{1/2}}{[(2L+3)(L+1)(2L+1)(2L-1)L]^{1/2}} \sum_M [3M^2 - L(L+1)]Q(M)$$

$$(4.6.11)$$

where explicit values for the $3j$ symbols have been used. In particular, if $L = 1$, the atomic system under consideration is completely specified by two parameters, the monopole or the total cross section and the alignment parameter $\langle T(1)_{20}^{\dagger} \rangle$. No net angular momentum $\langle \mathbf{L} \rangle$ is transferred to the system.

4.7. Time Evolution of State Multipoles in the Presence of an External Perturbation

4.7.1. The Perturbation Coefficients

The time evolution of state multipoles can be obtained from Eqs. (2.4.15) or (2.4.16) and (4.3.5). Of particular interest for our subsequent discussion is the following situation. Consider an ensemble of atoms (nuclei) which has been excited to states which can be described by a Hamiltonian $H = H_0 + H'$, where H' denotes a perturbation which is assumed to be weak and unimportant during the excitation process. The term H' can then be neglected during the excitation. Assuming that the eigenstates of H_0 can be identified in terms of angular momentum numbers we will denote the relevant states by $|JM\rangle$. After the excitation, however, the time evolution of the states is governed by the total Hamiltonian H and the corresponding time evolution operator $U(t)$.

Suppose that an atomic ensemble has been excited instantaneously at time $t = 0$ where by "instantaneously" is meant that the excitation time is much shorter than all characteristic transition times caused by the perturbation H' (see Section 3.5.1). Immediately after the excitation the ensemble can then be represented by a density matrix $\rho(0)$. Tensor operators can be constructed using the eigenstates $|JM\rangle$ of H_0 in Eqs. (4.2.8). Expanding $\rho(0)$

in terms of these tensors gives

$$\rho(0) = \sum_{\substack{J'J \\ KQ}} \langle T(J'J)_{KQ}^{\dagger} \rangle T(J'J)_{KQ} \qquad (4.7.1)$$

where the state multipoles are given by

$$\langle T(J'J)_{KQ}^{\dagger} \rangle = \operatorname{tr} \rho(0) T(J'J)_{KQ}^{\dagger} \qquad (4.7.2)$$

[see Eq. (4.3.5)]. We will henceforth denote the multipoles describing the atomic states at time $t = 0$ by $\langle T(J'J)_{KQ}^{\dagger} \rangle$. The sum in Eq. (4.7.1) includes all angular momenta J', J which are present at time $t = 0$.

The density matrix $\rho(0)$ evolves under the influence of the total Hamiltonian H into the density matrix

$$\rho(t) = U(t)\rho(0)U(t)^{\dagger} \qquad (4.7.3)$$

During the time interval $0 \cdots t$ the states which were initially excited are disturbed and mixed by the perturbation H'. Any state $|JM\rangle$ will evolve into a state $|\varphi(t)\rangle = U(t)|JM\rangle$ which can be expanded in terms of a full set of eigenstates $|jm\rangle$ of the Hamiltonian H_0. Tensor operators can be constructed from these states $|jm\rangle$ and used to expand $\rho(t)$:

$$\rho(t) = \sum_{\substack{j'j \\ kq}} \langle T(j'j)_{kq}^{\dagger} \rangle T(j'j)_{kq} \qquad (4.7.4)$$

where the sum includes all angular momenta j' and j in which the atoms can be found at time t. From Eqs. (4.7.2), (4.7.3), and (4.7.4)

$$\langle T(j'j)_{kq}^{\dagger} \rangle = \operatorname{tr} \rho(t) T(j'j)_{kq}^{\dagger}$$

$$= \operatorname{tr} U(t)\rho(0)U(t)^{\dagger} T(j'j)_{kq}^{\dagger}$$

$$= \sum_{\substack{J'J \\ KQ}} \langle T(J'J)_{KQ}^{\dagger} \rangle \operatorname{tr} [U(t)T(J'J)_{KQ}U(t)^{\dagger} T(j'j)_{kq}^{\dagger}]$$

$$= \sum_{\substack{J'J \\ KQ}} \langle T(J'J)_{KQ}^{\dagger} \rangle G(J'J, j'j; t)_{Kk}^{Qq} \qquad (4.7.5)$$

where we introduced the *perturbation coefficients*

$$\blacktriangleright \qquad G(J'J, j'j; t)_{Kk}^{Qq} = \operatorname{tr} [U(t)T(J'J)_{KQ}U(t)^{\dagger} T(j'j)_{kq}^{\dagger}] \qquad (4.7.6)$$

In Eq. (4.7.5) the multipoles $\langle T(j'j, t)_{kq}^{\dagger} \rangle$ characterizing the atomic states at time t, are expressed in terms of their counterparts at time $t = 0$. *The perturbation coefficients are simply the coefficients in this expansion.*

4.7.2. Perturbation Coefficients for the Fine and Hyperfine Interactions

In order to clarify the concepts introduced in the preceding section we will now consider the time evolution of atomic states, excited at $t = 0$, under the influence of the fine-structure interaction. We will not specify the excitation mechanism. The basic assumptions which will be made are that during the excitation the orbital and spin angular momentum of the atoms are uncoupled and that the atomic spins are unpolarized immediately after the excitation. With these assumptions the atomic states immediately after the excitation can then be represented in the uncoupled representation $|LMS_1M_{S_1}\rangle$ where the spin states are equally populated. We will assume sharp values of L and S_1.

These assumptions apply, for example, in the case of excitation of light atoms by electron impact as discussed in Section 3.5. They are also assumed to be valid for atoms which have been excited by beam–foil excitation.

Immediately after the excitation the atomic ensemble can be represented by a density matrix $\rho(0)$ with elements $\langle LM'S_1M'S_1|\rho(0)|LMS_1M_{S_1}\rangle$. In the following we will be interested only in the properties of the orbital states and hence we will define a reduced density matrix $\rho(L)$ with elements

$$\langle LM'|\rho(L)|LM\rangle = \sum_{M_{S_1}} \langle LM'S_1M_{S_1}|\rho(0)|LMS_1M_{S_1}\rangle \qquad (4.7.7)$$

For example, in the case of excitation by electron impact, the elements (4.7.7) are given by Eq. (3.5.7) in terms of the scattering amplitudes.

The reduced density matrix can be expanded in terms of state multipoles as in Eq. (4.6.1):

$$\rho(L) = \sum_{KQ} \langle T(L)_{KQ}^{\dagger}\rangle T(L)_{KQ} \qquad (4.7.8)$$

where the tensors $\langle T(L)_{KQ}\rangle$ characterize the orbital states at time $t = 0$, that is, immediately after the excitation.

The states $|LMS_1M_{S_1}\rangle$ can be considered as eigenstates of an Hamiltonian H_0. After the excitation the atoms are assumed to relax into the JM-coupling scheme under the influence of the fine-structure interaction $H' \sim LS$ which perturbs the initially excited states. In terms of the vector model this distortion can be interpreted as a precession of the angular momentum vectors $\mathbf{L}$ and $\mathbf{S}_1$ around the total angular momentum $\mathbf{J}$ (this precession is neglected during the excitation). The fine-structure interaction is assumed to be weak, and hence transitions between states with different L and S_1 can be neglected and L and S_1 can be considered to be conserved.

The time development of the atomic state is governed by the operator

$$U(t) = \exp(-iHt/\hbar)$$

where the total Hamiltonian $H = H_0 + H'$ includes the interaction term H' which couples the spin and orbital systems. Thus $U(t)$ acts on both systems and it is therefore necessary to consider the total density matrix $\rho(0)$ instead of the matrix (4.7.8), which only describes the orbital states. A convenient form of $\rho(0)$ can be obtained as follows: Since orbital and spin systems are uncorrelated at time $t = 0$ and the spins unpolarized, $\rho(0)$ is given by

$$\rho(0) = \rho(L) \times \frac{1}{2S_1 + 1}\mathbf{1} \tag{4.7.9a}$$

where Eqs. (2.2.14) and (A11) have been used and where $\mathbf{1}$ is the unit operator in spin space. Substituting Eq. (4.7.8) into Eq. (4.7.9a) gives

$$\rho(0) = \frac{1}{2S_1 + 1} \sum_{KQ} \langle T(L)_{KQ}^\dagger \rangle [T(L)_{KQ} \times \mathbf{1}] \tag{4.7.9b}$$

At time t the system is represented by a density matrix which has evolved from the matrix $\rho(0)$ and satisfies Eq. (2.4.15):

$$\rho(t) = U(t)\rho(0)U(t)^\dagger$$

$$= \frac{1}{2S_1 + 1} \sum_{KQ} \langle T(L)_{KQ}^\dagger \rangle U(t)(T(L)_{KQ} \times \mathbf{1})U(t)^\dagger \tag{4.7.10}$$

We define state multipoles $\langle T(Lt)_{kq}^\dagger \rangle$ describing the orbital states at time t as the irreducible components of the corresponding reduced density matrix $\rho(L, t)$:

$$\langle T(L, t)_{kq}^\dagger \rangle = \operatorname{tr} \rho(L, t)T(L)_{kq}^\dagger \tag{4.7.11a}$$

where the elements of $\rho(L, t)$ are given by

$$\langle LM'|\rho(L, t)|LM \rangle = \sum_{M_{S_1}} \langle LM'S_1M_{S_1}|\rho(t)|LMS_1M_{S_1} \rangle$$

Alternatively, and more conveniently for the present discussion, the multipole components can be represented by

$$\langle T(L, t)_{kq}^\dagger \rangle = \operatorname{tr} \rho(t)[T(L)_{kq}^\dagger \times \mathbf{1}] \tag{4.7.11b}$$

where $\mathbf{1}$ is the unit operator in spin space. The equivalence of Eqs. (4.7.11a) and (4.7.11b) can be shown by calculating the trace (4.7.11) using the uncoupled states $|LMS_1M_{S_1}\rangle$ (see Appendix A).

Substituting Eq. (4.7.10) into Eq. (4.7.11) gives

$$
\langle T(L, t)_{kq}^{\dagger}\rangle = \frac{1}{2S_1 + 1} \sum_{KQ} \langle T(L)_{KQ}^{\dagger}\rangle \, \text{tr} \, \{U(t)
$$

$$
\times [T(L)_{KQ} \times \mathbf{1}]U(t)^{\dagger}[T(L)_{kq}^{\dagger} \times \mathbf{1}]\}
$$

$$
= \sum_{KQ} \langle T(L)_{KQ}^{\dagger}\rangle G(L; t)_{Kk}^{Qq} \qquad (4.7.12)
$$

where the perturbation coefficients $G(Lt)_{Kk}^{Qq}$ are the coefficients in this new multipole expansion:

$$
G(L; t)_{Kk}^{Qq} = \frac{1}{2S_1 + 1} \, \text{tr} \, \{U(t)[T(L)_{KQ} \times \mathbf{1}]U(t)^{\dagger}[T(L)_{kq}^{\dagger} \times \mathbf{1}]\} \quad (4.7.13)
$$

We will now derive an explicit expression for the quantities (4.7.13). Because the elements of $U(t)$ are diagonal in the eigenstate representation $|(LS_1)JM\rangle$ of the total Hamiltonian H the matrix representation of $U(t)$ in this representation has elements

$$
\langle (LS_1)J'M'|U(t)|(LS_1)JM\rangle = \exp{(-iE_Jt/h)}\delta_{JJ'}\cdot\delta_{MM'} \qquad (4.7.14)
$$

(where E_J is the energy of the level LS_1J). Using Eq. (4.7.14) the trace in Eq. (4.7.13) can be calculated in the coupled representation:

$$
G(L; t)_{Kk}^{Qq} = \frac{1}{2S_1 + 1} \sum_{\substack{J'M' \\ JM}} \exp{\left[\frac{+i(E_J - E_{J'})t}{\hbar}\right]}
$$

$$
\times \langle (LS_1)J'M'|T(L)_{KQ} \times \mathbf{1}|(LS_1)JM\rangle\langle(LS_1)JM|T(L)_{kq}^{\dagger} \times \mathbf{1}|(LS_1)J'M'\rangle
$$

$$
(4.7.15a)
$$

The remaining elements can be obtained by using the fact that $T(L)_{KQ} \times \mathbf{1}$ is a tensor operator of rank K and components Q. Applying the Wigner–Eckart theorem

$$
\langle (LS_1)J'M'|T(L)_{KQ} \times \mathbf{1}|(LS_1)JM\rangle
$$

$$
= (-1)^{J'-M'}\begin{pmatrix} J' & K & J \\ -M' & Q & M \end{pmatrix}\langle(LS_1)J'\|T_K \times \mathbf{1}\|(LS_1)J\rangle
$$

$$
(4.7.15b)
$$

and a standard formula of angular momentum theory [Eq. (C20)] gives

$$
\langle (LS_1)J'\|T_K \times \mathbf{1}\|(LS_1)J\rangle
$$

$$
= (-1)^{L+S_1+J+K}[(2J' + 1)(2J + 1)(2K + 1)]^{1/2}\begin{Bmatrix} L & J' & S_1 \\ J & L & K \end{Bmatrix}
$$

$$
(4.7.15c)
$$

where Eq. (4.2.28) has been used to give the explicit value of the reduced matrix element $\langle L \| T(L)_K \| L \rangle$. A similar formula holds for the matrix element of the operator $T(L)_{kq}^\dagger \times \mathbf{1}$. Inserting the resultant expressions into Eq. (4.7.15a) and summing over M' and M with the help of the orthogonality relations of the $3j$ symbols gives

$$G(L;t)_{Kk}^{Qq} = \frac{1}{2S_1 + 1} \sum_{J'J} (2J' + 1)(2J + 1)$$
$$\times \begin{Bmatrix} L & J' & S_1 \\ J & L & K \end{Bmatrix}^2 \exp\left[\frac{-i(E_{J'} - E_J)t}{\hbar}\right]\delta_{Kk}\delta_{Qq} \quad (4.7.15)$$

The Kronecker symbols indicate that multipoles with different ranks and components cannot be mixed by the interaction. Furthermore, the perturbation coefficients are independent of Q and can hence be written in the form

$$G(L;t)_{Kk}^{Qq} = G(L;t)_K \delta_{Kk}\delta_{Qq} \quad (4.7.16)$$

The coefficients (4.7.15) are real numbers. This can be shown by taking the complex conjugate of the bracket in Eq. (4.7.15), interchanging J and J', and using the symmetry property Eq. (C8) of the $6j$ symbols. The imaginary part of the complex exponential function in Eq. (4.7.15) therefore vanishes as a result of summing over all J and J' and only the real part survives:

▶ $$G(L;t)_K = \frac{1}{2S_1 + 1} \sum_{J'J} (2J' + 1)(2J + 1)\begin{Bmatrix} L & J' & S_1 \\ J & L & K \end{Bmatrix}^2 \cos\left[\frac{(E_{J'} - E_J)t}{\hbar}\right]$$
$$(4.7.17)$$

From Eqs. (4.7.12) and (4.7.16) we obtain finally

▶ $$\langle T(L;t)_{KQ}^\dagger \rangle = G(L;t)_K \langle T(L)_{KQ}^\dagger \rangle \quad (4.7.18)$$

which describes the time evolution of the state multipoles under the influence of the fine-structure interaction.

It is sometimes convenient to represent the coefficients $G(L;t)_K$ in the form

$$G(L;t)_K = \overline{G(L)_K} + \frac{1}{2S_1 + 1} \sum_{J' \neq J} (2J' + 1)(2J + 1)$$
$$\times \begin{Bmatrix} L & J' & S_1 \\ J & L & K \end{Bmatrix}^2 \cos\left[\frac{(E_{J'} - E_J)t}{\hbar}\right] \quad (4.7.19)$$

where terms with $J = J'$ and terms with $J \neq J'$ have been separated and where the time-independent part is defined by

$$\overline{G(L)_K} = \frac{1}{2S_1 + 1} \sum_J (2J + 1)^2 \begin{Bmatrix} L & J & S_1 \\ J & L & K \end{Bmatrix}^2 \quad (4.7.20)$$

The hyperfine interaction can be treated by using the same method as has been used for the fine-structure interaction. Suppose that at time $t = 0$ atomic states with electronic angular momentum J are excited with the nuclear spin I unaffected. Constructing state multipoles $\langle T(J)_{KQ}^{\dagger} \rangle$ and $\langle T(J;t)_{KQ}^{\dagger} \rangle$ from the states $|JM\rangle$ at times $t = 0$ and t, respectively, it can be shown that these tensors are related by an expression similar to Eq. (4.6.18):

$$\langle T(J;t)_{KQ}^{\dagger} \rangle = G(J;t)_K \langle T(J)_{KQ}^{\dagger} \rangle$$

where the perturbation coefficients $G(J;t)$ are given by Eq. (4.7.17) with L replaced by J, S_1 replaced by I, and $J(J')$ replaced by $F(F')$, where F denotes the total angular momentum:

$$G(J;t)_K = \frac{1}{2I+1} \sum_{F'F} (2F'+1)(2F+1) \begin{Bmatrix} J & F' & I \\ F & J & K \end{Bmatrix}^2 \cos\left[\frac{(E_{F'}-E_F)t}{\hbar}\right]$$

$$(4.7.21)$$

Finally, we will consider the case in which both fine and hyperfine interaction must be taken into account. When the hyperfine interaction is much weaker than the fine-structure interaction the angular momentum of the electronic state, J, remains a good quantum number and the relevant perturbation coefficients can be calculated by a similar method to that used above and are given by

$$G(t)_K^H = \frac{1}{(2S_1+1)(2I+1)} \sum_{\substack{J'J \\ F'F}} (2J+1)(2J'+1)(2F'+1)$$

$$\times (2F+1) \begin{Bmatrix} J' & F' & I \\ F & J & K \end{Bmatrix}^2 \begin{Bmatrix} L & J' & S_1 \\ J & L & K \end{Bmatrix}^2$$

$$\times \exp\left[\frac{i(E_{1'}-E_1)t}{\hbar} - \frac{(\gamma_1+\gamma_{1'})t}{2}\right] \qquad (4.7.22)$$

In deriving Eq. (4.7.22) it is again assumed that S_1 and I are unaffected by the excitation and decay process. The energies E_1 and decay constants γ_1 refer to states with angular momenta J and F.

If the fine and hyperfine splittings are comparable a more elaborate calculation is necessary since J is no longer a good quantum number. This is discussed in more detail, for example, by Fano and Macek (1973).

4.7.3. An Explicit Example

We will now discuss the physical significance of the perturbation coefficients (4.7.17), following Fano and Macek (1973), using an explicit example. Let us take the case with $L = 1$, $S_1 = 1/2$, and $J = 1/2$, $3/2$, and discuss the time evolution of the orientation vector:

$$\langle T(L;t)_{1Q}^{\dagger}\rangle = G(L;t)_1\langle T(L)_{1Q}^{\dagger}\rangle \qquad (4.7.23)$$

The numerical values of the relevant $6j$ symbols are given by

$$\begin{Bmatrix} 1 & \frac{1}{2} & \frac{1}{2} \\ \frac{1}{2} & 1 & 1 \end{Bmatrix} = -\frac{1}{3}, \quad \begin{Bmatrix} 1 & \frac{3}{2} & \frac{1}{2} \\ \frac{3}{2} & 1 & 1 \end{Bmatrix} = \frac{5^{1/2}}{6(2^{1/2})},$$

$$\begin{Bmatrix} 1 & \frac{1}{2} & \frac{1}{2} \\ \frac{3}{2} & 1 & 1 \end{Bmatrix} = -\frac{1}{6}$$

Substituting these values into Eq. (4.7.17) gives

$$G(L;t)_1 = \tfrac{7}{9} + \tfrac{2}{9}\cos\left[(E_{1/2} - E_{3/2})t/\hbar\right] \qquad (4.7.24)$$

Equation (4.7.24) shows that in this case $G(L;t)_1$ oscillates around the mean value $\overline{G(L)_1} = 7/9$ with a frequency $\omega = (E_{1/2} - E_{3/2})/\hbar$, which in the semiclassical model is just the precession period of the vectors S_1 and L around J. From Eqs. (4.7.23) and (4.7.24) it can be seen that the orientation vector $\langle T(L)_{1Q}^{\dagger}\rangle$ varies periodically with t. This behavior is a consequence of angular momentum coupling. During the excitation the orbital system aquires a certain orientation while the spins remain unpolarized. Because of the spin–orbit coupling, which is assumed to be "switched on" immediately after the excitation, there is a transfer of orientation between orbital and spin systems. The spins become oriented and there is a resultant loss of orientation in the orbital states. In each period $\langle T(L)_{1Q}^{\dagger}\rangle$ decreases, reaches a minimum (when the spins have obtained the maximum possible orientation), and then rises again to its initial value when the spins are again unoriented. *This exchange of orientation is periodic and reversible*, reflecting the fact that the spin–orbit coupling $H' \sim LS$ is *symmetric* in L and S. These results can be generalized to any multipole component.

In conclusion, the state multipoles $\langle T(Lt)_{KQ}^{\dagger}\rangle$ oscillate about a mean value $\overline{G(L)_K}\langle T(L)_{KQ}^{\dagger}\rangle$. This behavior is a consequence of the fine-structure interaction between orbital and spin angular momentum which results in a *periodic and reversible* exchange of polarization between the two systems. The observable consequences of these time variations will be considered in detail in Chapters 5 and 6.

4.7.4. Influence of an External Magnetic Field

We will now consider a situation in which an ensemble of atoms (or nuclei) has been excited at time $t = 0$ in the presence of a magnetic field, **B**. The total Hamiltonian is given by $H = H_0 + H'$, where $H' = -g\mu_B \mathbf{JB}$

describes the interaction with the field. The eigenstates of H_0 are chosen to be $|JM\rangle$. The magnetic field induces splittings between levels with the same J and different M. Transitions between the split levels are characterized by a transition time $T \sim 1/\Delta E$, where ΔE is the largest energy splitting of a level with angular momentum J. It will be assumed that T is large compared to the excitation time. In this case the influence of the magnetic field can be neglected during the excitation and the atoms can be assumed to be excited into states $|JM\rangle$. The field will disturb the states immediately after the excitation and we will consider the time development of the initial multipoles $\langle T(J)_{KQ}^{\dagger}\rangle$ under the influence of the time evolution operator

$$U(t) = \exp\left(-iHt\right)/\hbar \qquad (4.7.25)$$

The relevant perturbation coefficients are given by

$$G(J;t)_{Kk}^{Qq} = \text{tr}\left[U(t)T(J)_{KQ}U(t)^{\dagger}T(J)_{kq}^{\dagger}\right] \qquad (4.7.26)$$

The initial multipoles are defined with respect to a coordinate system XYZ and the field **B** is taken to be parallel to a direction z. With z as the quantization axis

$$H' = -g\mu_B J_z B \qquad (4.7.27)$$

and relating the eigenstates of H_0 to z the elements of H are given by

$$\langle Jm'|H|Jm\rangle = (E_J - g\mu_B Bm)\delta_{m'm} \qquad (4.7.28)$$

Transforming the tensor operators to the xyz system using Eq. (4.2.13), calculating the trace (4.7.26) in terms of the states $|JM\rangle$, and using Eqs. (4.7.25) and (4.7.28) gives, after some manipulations,

$$\begin{aligned}
G(J;t)_{Kk}^{Qq} &= \sum_{Q'q'} (-1)^q [D(0\beta\alpha)_{Q'Q}^{(K)}][D(0\beta\alpha)_{q'-q}^{(k)}] \\
&\quad \times \text{tr}\left[U(t)T(J)_{KQ'}U(t)^{\dagger}T(J)_{kq'}\right] \\
&= \sum_{Q'q'} (-1)^q [D(0\beta\alpha)_{Q'Q}^{(K)}][D(0\beta\alpha)_{q'-q}^{(k)}] \\
&\quad \times \sum_{m'm} \langle Jm'|T(J)_{KQ'}|Jm\rangle \\
&\quad \times \langle Jm|T(J)_{kq'}|Jm'\rangle \exp\left(-i\omega_L Q't\right)
\end{aligned} \qquad (4.7.29)$$

where β is the angle between Z and z and α is the azimuthal angle of z in the XYZ system. ω_L denotes the Larmor frequency.

Using Eq. (4.2.9) in the xyz system, performing the sum over m' and m by using the orthogonality relations of the $3j$ symbols, we obtain

$$G(J;t)_{Kk}^{Qq} = \delta_{Kk} \sum_{Q'} \exp\left(-i\omega_L Q't\right) D(0\beta\alpha)_{Q'Q}^{(K)} D(0\beta\alpha)_{Q'q}^{(K)*} \qquad (4.7.30)$$

Using Eq. (C12), $\exp(-i\omega_L Qt)$ can be interpreted as a rotation of $-\omega_L t$ about Z.

In the simple case where the field direction z coincides with Z we have $\beta = \alpha = 0$ and Eq. (4.7.30) reduces to

$$G(J,t)_{Kk}^{Qq} = \delta_{Kk}\delta_{Qq}\exp(-i\omega_L Qt) \qquad (4.7.31)$$

In this case the field cannot alter the multipoles but merely causes their phases to change in time:

$$\langle T(J;t)_{KQ}^{\dagger}\rangle = \exp(-i\omega_L Qt)\langle T(J)_{KQ}^{\dagger}\rangle \qquad (4.7.32)$$

Equations (4.7.30) and (4.7.31) will be used in Section 6.3.

4.8. Notations Used by Other Authors

Unfortunately many notations are used for the tensor operators and state multipoles. We list here a few of the notations used in the literature. Our tensor operator $T(J'J)_{KQ}$ is written as

$T_Q^{(K)}(\alpha J', \beta J)$ by Omont (1977),

$T_Q^{(K)}(J', J)$ by Happer (1972), and

T_{KQ} by Brink and Satchler (1962) and Lamb and Ter Haar (1971).

Dyanokov and Perel (1965) use a different normalization so that the operator $T_Q^{(K)}$ of these authors is equal to

$$[(2K+1)/(2J+1)]^{1/2}(-1)^Q T(J)_{KQ}$$

in our notation.

Our state multipole $\langle T(J'J)_{KQ}^{\dagger}\rangle$ is written as

$\rho_Q^{(K)}(\alpha J', \beta J)$ by Omont (1977),

$\rho_{KQ}(J'J)$ by Happer (1972),

$\rho(JJ')_{KQ}^*$ by Brink and Satchler (1962) and Lamb and Ter Haar (1971),

$(-1)^{J'+J+Q}\rho_Q^{(K)}(J'J)$ by Steffen and Alder (1975).

5
Radiation from Polarized Atoms. Quantum Beats

5.1. General Theory I: Density Matrix Description of Radiative Decay Processes

In this chapter we will consider the decay of an ensemble of excited atoms by photon emission. We will discuss the following case. We assume that an ensemble of atoms has been excited "instantaneously" at time $t = 0$. As usual, "instantaneous" implies that the excitation time is short compared to the mean lifetime of the excited states and any characteristic precession frequency (see Sections 3.5 and 4.7.1).

The excitation mechanism can be of any kind and the atoms may have been excited, for example, by electron impact, photon absorption, or by beam–foil techniques. Our main objective will be the derivation of Eqs. (5.2.6) and (5.2.7), which will then be used in the subsequent sections. Readers who are not so much interested in mathematical details of the proofs may proceed directly to formulas (5.2.6) and (5.2.7).

The excited atoms may be considered as a (coherent or incoherent) superposition of states $|\alpha_1 J_1 M_1\rangle$, where α_1 denotes collectively the set of quantum numbers which are needed to describe the states in addition to the angular momentum quantum numbers J_1 and M_1.

The atoms are assumed to decay to lower levels $|\alpha_2 J_2 M_2\rangle$. In the following we will assume α_1 and α_2 to be fixed and suppress the dependence of the state vectors on these quantum numbers. The treatment will be further simplified by neglecting the finite lifetime of the final state.

The emitted photons will be described in terms of state vectors $|\mathbf{n}\omega\lambda\rangle$ in the helicity representation, where $\mathbf{n}$ is fixed by the direction of observation. We will now derive an expression for the polarization density matrix of the emitted photons. In the first part of the calculation we will use $\mathbf{n}$ as the quantization axis and relate all angular momentum quantum numbers to this axis. This choice will considerably simplify the second part of the calculations in the following section. At the end of the calculations we will transform to a coordinate system defined with respect to the excitation process.

Immediately after the excitation the ensemble of excited atoms may be characterized by a density matrix $\rho(0)$ which evolves according to Eq. (2.4.15) into a density matrix:

$$\rho(t)_{\text{out}} = U(t)\rho(0)U(t)^{\dagger} \tag{5.1.1}$$

where the operator $U(t)$ describes the time evolution under the influence of the interaction with the virtual radiation field. (In this section it is assumed that the atomic states are not perturbed by internal or external fields between excitation and decay.) The matrix ρ_{out} describes the entire ensemble of atoms and photon at time t, that is, the atoms which are still in the excited state, the atoms in the lower levels, and the photons emitted in the time interval $0 \cdots t$.

The decay process can be described in first-order perturbation theory. In this approximation the operator $U(t)$ is given by Eqs. (2.4.35) and (2.4.31):

$$U(t) = U(t)_0\left[1 - \frac{i}{\hbar}\int_0^t V(\tau)_I\, d\tau\right]$$

$$= U(t)_0\left[1 - \frac{i}{\hbar}\int_0^t U(\tau)_0^{\dagger}VU(\tau)_0\, d\tau\right] \tag{5.1.2}$$

where we inserted Eq. (2.4.25). The elements of the operator V, describing the interaction between atoms and virtual radiation field, will be specified later. $U(t)_0$ is the free time evolutions operator; hence

$$U(t)_0|J_2M_2\omega\mathbf{n}\lambda\rangle = \exp\left[-(i/\hbar)E_2t - i\omega t\right]|J_2M_2\omega\mathbf{n}\lambda\rangle \tag{5.1.3a}$$

$$U(t)_0|J_1M_1\rangle = \exp\left[-(i/\hbar)E_1t - \gamma_1t/2\right]|J_1M_1\rangle \tag{5.1.3b}$$

where the finite level width of the initial state has been included. E_1 and E_2 denote the energy of the states with angular momentum J_1 and J_2, respectively, and γ_1 is the decay constant.

We are interested in the elements $\langle\omega\mathbf{n}\lambda'|\rho(t)|\omega\mathbf{n}\lambda\rangle$ of the reduced density matrix $\rho(t)$ which describe the polarization state of the emitted photons only. The normalization is as in Eq. (1.2.17) so that the diagonal elements $\langle\omega\mathbf{n}\lambda|\rho(t)|\omega\mathbf{n}\lambda\rangle$ give the intensity of those photons detected in the direction $\mathbf{n}$ with frequency ω and helicity λ. Using Eq. (3.2.3) the matrix elements are given by

$$\langle \omega \mathbf{n}\lambda'|\rho(t)|\omega \mathbf{n}\lambda \rangle = \sum_{J_2 M_2} \langle J_2 M_2 \omega \mathbf{n}\lambda'|\rho(t)_{\text{out}}|J_2 M_2 \omega \mathbf{n}\lambda \rangle \qquad (5.1.4)$$

Substituting Eq. (5.1.1) for $\rho(t)_{\text{out}}$ and Eq. (5.1.2) for $U(t)$ gives‡

$$\langle \omega \mathbf{n}\lambda'|\rho(t)|\omega \mathbf{n}\lambda \rangle = \frac{\hbar\omega}{\hbar^2} \sum_{\substack{J_2 M_2 \\ J\{M\{J_1 M_1}} \langle J_2 M_2 \omega \mathbf{n}\lambda'| \int_0^t U(\tau)_0^{\dagger} V U_0(\tau)\, d\tau|J_1' M_1' \rangle$$

$$\times \langle J_1' M_1'|\rho(0)|J_1 M_1\rangle$$

$$\times \langle J_1 M_1|\int_0^t U(\tau)_0^{\dagger} V U(\tau)_0\, d\tau|J_2 M_2 \omega \mathbf{n}\lambda \rangle$$

The terms proportional to the identity matrix cannot contribute to the transitions $J_1 \rightarrow J_2 \neq J_1$. By applying Eq. (5.1.3) we obtain

$$\langle \omega \mathbf{n}\lambda'|\rho(t)|\omega \mathbf{n}\lambda \rangle = \frac{\omega}{\hbar} \sum_{\substack{J_2 M_2 \\ J\{M\{J_1 M_1}} \langle J_2 M_2 \omega \mathbf{n}\lambda'|V|J_1' M_1'\rangle\langle J_1' M_1'|\rho(0)|J_1 M_1\rangle$$

$$\times \langle J_1 M_1|V|J_2 M_2 \omega \mathbf{n}\lambda\rangle \left\{\left[\int_0^t d\tau \exp\left[i\left(\omega + \omega_{21'} + \frac{i}{2}\gamma_1'\right)\tau\right]\right]\right\}$$

$$\times \left\{\left[\int_0^t d\tau \exp\left[i\left(-\omega - \omega_{21} + \frac{i\gamma_1}{2}\right)\tau\right]\right]\right\} \qquad (5.1.5)$$

with $\omega_{21'} = (1/\hbar)(E_2 - E_{1'})$, $\omega_{21} = (E_2 - E_1)/\hbar$, and where $E_{1'}$, $\gamma_{1'}$ and E_1, γ_1 denote energy and decay constants of the states $|J_1' M_1'\rangle$ and $|J_1 M_1\rangle$, respectively.

Taking the nonrelativistic limit the elements of V are given in the dipole approximation by

$$\langle J_2 M_2 \omega \mathbf{n}\lambda|V|J_1 M_1\rangle = -i\omega_{21}[(2\pi\hbar/\omega)]^{1/2} e\langle J_2 M_2|\mathbf{e}_\lambda^* \mathbf{r}|J_1 M_1\rangle \qquad (5.1.6)$$

(for details, see, for example, Landau and Lifschitz, 1970), where $\mathbf{e}_\lambda$ denotes the polarization vector (1.2.8) ($\lambda = \pm 1$) and $\mathbf{r}$ is the dipole operator. The time integrations in Eq. (5.1.5) can easily be carried out:

$$\langle \omega \mathbf{n}\lambda'|\rho(t)|\omega \mathbf{n}\lambda \rangle = \frac{2\pi e^2 \omega_{21}^2}{\omega \hbar^2} \sum_{\substack{J_2 M_2 \\ J\{M\{J_1 M_1}} \langle J_2 M_2|\mathbf{e}_\lambda^* \mathbf{r}|J_1' M_1'\rangle\langle J_1' M_1'|\rho(0)|J_1 M_1\rangle$$

$$\times \langle J_1 M_1|(\mathbf{e}_\lambda^* \mathbf{r})^{\dagger}|J_2 M_2\rangle\left(\frac{\exp\left[i(\omega + \omega_{21'} + i\gamma_{1'}/2)t\right] - 1}{i(\omega + \omega_{21'} + i\gamma_{1'}/2)}\right)$$

$$\times \left(\frac{\exp\left[i(\omega + \omega_{21} + i\gamma_1/2)t\right] - 1}{i(\omega + \omega_{21} + i\gamma_1/2)}\right) \qquad (5.1.7)$$

‡ We have to multiply by $\hbar\omega$ in accordance with the normalization (1.2.17).

In obtaining the numerical factor multiplying the right-hand side we have set $\omega_{21} \approx \omega_{21'}$ since the splitting of the upper levels is much smaller than the energy difference between upper and lower states. Finally, we multiply with the density of final states $\omega^2 \, d\Omega \, d\omega/(2\pi c)^3 \hbar$ and integrate both sides of Eq. (5.1.7) over the line profile. Because the main contributions come from the region $\omega \approx \omega_{21}$ the integral over ω can be extended to $-\infty$ with negligible error and the ω integration can then be carried out in the complex ω plane by using Cauchy's integral formula. This gives

$$\rho(\mathbf{n}, t)_{\lambda'\lambda} = C(\omega) \sum_{\substack{J_2 M_2 \\ J_1' M_1' J_1 M_1}} \langle J_2 M_2 | \mathbf{e}_\lambda^* \mathbf{r} | J_1' M_1' \rangle \langle J_1' M_1' | \rho(0) | J_1 M_1 \rangle$$

$$\times \langle J_1 M_1 | (\mathbf{e}_\lambda^* \mathbf{r})^\dagger | J_2 M_2 \rangle$$

$$\times \frac{1 - \exp\left[-i(E_{1'} - E_1)t/\hbar - (\gamma_1 + \gamma_{1'})t/2\right]}{i(E_{1'} - E_1)/\hbar + (\gamma_1 + \gamma_{1'})/2} \tag{5.1.8}$$

where

$$C(\omega) = \frac{e^2 \omega^4 \, d\Omega}{2\pi c^3 \hbar} \tag{5.1.9}$$

and where $d\Omega$ is the element of solid angle into which the photons are emitted. In Eq. (5.1.8) the notation $\rho(\mathbf{n}, t)_{\lambda'\lambda}$ is introduced for the elements of the obtained density matrix. The polarization vectors $e_{\lambda'}$ and e_λ in Eq. (5.1.9) can be eliminated by noting that in the helicity system the coordinate system is "spanned" by the three unit vectors $\mathbf{e}_{+1}$, $\mathbf{e}_{-1}$, $\mathbf{n}$, and that the dipole vector $\mathbf{r}$ can be expanded in terms of this basis:

$$\mathbf{r} = r_{+1}^* \mathbf{e}_{+1} + r_{-1}^* \mathbf{e}_{-1} + r_0^* \mathbf{n}$$

where $r_{\pm 1}$ and r_0 are the components of $\mathbf{r}$ along the directions of $\mathbf{e}_{\pm 1}$ and $\mathbf{n}$, respectively. That is, $r_{\pm 1}$ and r_0 are therefore the spherical components of the vector $\mathbf{r}$. In this system the scalar product of $\mathbf{r}$ and e_λ is given by

$$\mathbf{e}_\lambda^* \cdot \mathbf{r} = r_\lambda^* = -r_{-\lambda} \tag{5.1.10}$$

This then finally gives the elements of the polarization density matrix of photons observed in the direction $\mathbf{n}$:

$$\rho(\mathbf{n}, t)_{\lambda'\lambda} = C(\omega) \sum_{\substack{J_2 M_2 \\ J_1' M_1' J_1 M_1}} \langle J_2 M_2 | r_{-\lambda'} | J_1' M_1' \rangle$$

$$\times \langle J_1' M_1' | \rho(0) | J_1 M_1 \rangle \langle J_1 M_1 | r_{-\lambda}^\dagger | J_2 M_2 \rangle$$

$$\times \frac{1 - \exp\left[-i(E_{1'} - E_1)t/\hbar - (\gamma_{1'} + \gamma_1)t/2\right]}{i(E_{1'} - E_1)/\hbar + (\gamma_1 + \gamma_{1'})/2} \tag{5.1.11}$$

5.2. General Theory II: Separation of Dynamical and Geometrical Factors

In order to disentangle the dynamical and geometrical factors in Eq. (5.1.11) and to take angular momentum conservation explicitly into account we can apply the irreducible tensor operator method. First of all $\rho(0)$ is expanded in terms of state multipoles, characterizing the excited atoms immediately after the excitation defined in the system with **n** as the quantization axis:

$$\rho(0) = \sum_{KqJ_1'J_1} \langle T(J_1'J_1)_{Kq}^\dagger \rangle T(J_1'J_1)_{Kq} \tag{5.2.1}$$

Substitution of Eq. (5.2.1) into Eq. (5.1.11) yields

$$\rho(\mathbf{n}, t)_{\lambda'\lambda} = C(\omega) \sum_{KqJ_1'J_1} \text{tr}\left[r_{-\lambda'}T(J_1'J_1)_{Kq}r_{-\lambda}^\dagger\right]\langle T(J_1'J_1)_{Kq}^\dagger \rangle$$

$$\times \frac{1 - \exp\left[-i(E_{1'} - E_1)t/\hbar - (\gamma_1 + \gamma_{1'})t/2\right]}{i(E_{1'} - E_1)/\hbar + (\gamma_{1'} + \gamma_1)/2} \tag{5.2.2}$$

where the trace is given by the sum

$$\text{tr}\left[r_{-\lambda'}T(J_1'J_1)_{Kq}r_{-\lambda}^\dagger\right] = \sum_{J_2M_2M_1M_1'} \langle J_2M_2|r_{-\lambda'}|J_1'M_1'\rangle$$

$$\times \langle J_1'M_1'|T(J_1'J_1)_{Kq}|J_1M_1\rangle\langle J_1M_1|r_{-\lambda}^\dagger|J_2M_2\rangle \tag{5.2.3}$$

In order to perform the sum in Eq. (5.2.3) we first apply the Wigner–Eckart theorem (4.2.27), which allows the dynamical factors (reduced matrix elements) and the geometrical elements ($3j$ symbols) to be separated, and then we use Eq. (C9) (Appendix C) in order to express the sum over $M_1'M_1M_2$ in terms of a $6j$ symbol. Inserting the elements of the tensor operators (4.2.9) we obtain

$$\text{tr}\left(r_{-\lambda'}T(J_1'J_1)_{Kq}r_{-\lambda}^\dagger\right) = \sum_{J_2} \langle J_2\|\mathbf{r}\|J_1'\rangle\langle J_2\|\mathbf{r}\|J_1\rangle^*(-1)^{J_1+J_2+\lambda}(2K+1)^{1/2}$$

$$\times \begin{pmatrix} 1 & 1 & K \\ -\lambda' & \lambda & q \end{pmatrix}\begin{Bmatrix} 1 & 1 & K \\ J_1 & J_1' & J_2 \end{Bmatrix} \tag{5.2.4}$$

While the multipoles $\langle T(J_1'J_1)_{Kq}^\dagger \rangle$ are defined with respect to **n** as the quantization axis, the information on the excited states is usually given in terms of tensors $\langle T(J_1'J_1)_{KQ}^\dagger \rangle$ which are defined in a coordinate system XYZ which is more appropriate to a description of the excitation process (for example, the "collision system" introduced in Section 3.5). As the last step in our calculations, we therefore transform the tensors from the helicity to the XYZ system. If θ and φ are the polar angles of $\mathbf{n}$ in the XYZ system (see, for example, Figure 4.1, where the axis z may now denote the direction of $\mathbf{n}$), then Eq. (4.3.13) gives

$$\langle T(J_1'J_1)_{KQ}^\dagger \rangle = \sum_q \langle T(J_1'J_1)_{Kq}^\dagger \rangle D(0\theta\varphi)_{qQ}^{(K)*} \qquad (5.2.5a)$$

with the inverse relation

$$\langle T(J_1'J_1)_{Kq}^\dagger \rangle = \sum_Q \langle T(J_1'J_1)_{KQ}^\dagger \rangle D(0\theta\varphi)_{qQ}^{(K)} \qquad (5.2.5)$$

Substitution of Eqs. (5.2.4) and (5.2.5) into Eq. (5.2.2) then finally yields

▶ $$\rho(n,t)_{\lambda'\lambda} = C(\omega) \sum_{J_2J_1'J_1KQq} \langle J_2\|\mathbf{r}\|J_1'\rangle \langle J_2\|\mathbf{r}\|J_1\rangle^* (-1)^{J_1+J_2+\lambda}(2K+1)^{1/2}$$

$$\times \begin{pmatrix} 1 & 1 & K \\ -\lambda' & \lambda & q \end{pmatrix} \begin{Bmatrix} 1 & 1 & K \\ J_1 & J_1' & J_2 \end{Bmatrix} D(0\theta\varphi)_{qQ}^{(K)} \langle T(J_1'J_1)_{KQ}^\dagger \rangle$$

$$\times \frac{1 - \exp[-i(E_{1'}-E_1)t/\hbar - (\gamma_{1'}+\gamma_1)t/2]}{i(E_{1'}-E_1)/\hbar + (\gamma_{1'}+\gamma_1)/2} \qquad (5.2.6)$$

with $q = \lambda' - \lambda$. Note that the helicity is invariant with respect to rotation so that λ' and λ have the same values in both coordinate systems.

Equation (5.2.6) gives the polarization density matrix of photons observed in the direction $\mathbf{n}$ and emitted in the time interval $0\cdots t$. One may also determine the state of those photons which are only emitted at time t (that is, in a short time interval $t\cdots t + dt$). The relevant density matrix is obtained by differentiating Eq. (6.2.6) with respect to the time. Denoting the time derivative of the density matrix by $\dot\rho(\mathbf{n},t)_{\lambda'\lambda}$ we obtain

▶ $$\dot\rho(\mathbf{n},t)_{\lambda'\lambda} = C(\omega) \sum_{J_1'J_1J_2KQq} \langle J_2\|\mathbf{r}\|J_1'\rangle \langle J_2\|\mathbf{r}\|J_1\rangle^* (-1)^{J_1+J_2+\lambda}(2K+1)^{1/2}$$

$$\times \begin{pmatrix} 1 & 1 & K \\ -\lambda' & \lambda & q \end{pmatrix} \begin{Bmatrix} 1 & 1 & K \\ J_1 & J_1' & J_2 \end{Bmatrix} D(0\theta\varphi)_{qQ}^{(K)}$$

$$\times \exp\left[\frac{-i(E_{1'}-E_1)t}{\hbar} - \frac{(\gamma_{1'}+\gamma_1)t}{2}\right] \langle T(J_1'J_1)_{KQ}^\dagger \rangle \qquad (5.2.7)$$

The density matrix elements can then be further expressed in terms of the Stokes parameters by means of Eq. (1.2.24). From this expression and from Eq. (1.2.29) all information on the behavior of the radiation in polarization experiments can be obtained.

5.3. Discussion of the General Formulas

5.3.1. General Structure of the Equations

Equations (5.2.6) and (5.2.7) will form the basis of our discussions in the remainder of this and the following chapter. It is therefore useful to discuss their meaning in more detail. We will concentrate here on Eq. (5.2.7) and postpone a discussion of Eq. (5.2.6) until Section 5.5, although many of the following remarks apply in both cases.

Equation (5.2.7) is the polarization density matrix of those photons emitted at time t. The time evolution of the initially excited atomic states between the excitation and decay is assumed to be undisturbed except by the decay process and is characterized by the time evolution operator

$$U(t)_0 = \exp\left[-iH_0 t/\hbar - \Gamma t/2\right] \tag{5.3.1}$$

where the states $|J_1 M_1\rangle$ are eigenfunctions of H_0 and where Γ denotes the decay matrix:

$$\exp\left(-\Gamma t/2\right)|J_1 M_1\rangle = \exp\left(-\gamma_1 t/2\right)|J_1 M_1\rangle$$

By inserting Eq. (5.3.1) into Eq. (4.7.6) it may be shown that the corresponding perturbation coefficient is

$$G(J_1' J_1 j_1' j_1, t)_{Kk}^{Qq} = \exp\left[-i\omega_{1'1} t - (\gamma_{1'} + \gamma_1)t/2\right]\delta_{J_{ij}i}\delta_{J_{1j}j_1}\delta_{Kk}\delta_{Qq} \tag{5.3.2}$$

for all K and Q. The state multipoles describing the excited states at time t are given by

$$\langle T(J_1' J_1, t)_{KQ}^{\dagger}\rangle = \exp\left[-i\omega_{1'1} t - (\gamma_{1'} + \gamma_1)t/2\right]\langle T(J_1' J_1)_{KQ}^{\dagger}\rangle \tag{5.3.3}$$

Reading Eq. (5.2.7) from the right to the left and taking Eq. (5.3.2) into account we have the following scheme:

$$\dot{\rho}(\mathbf{n}, t)_{\lambda'\lambda} = [\cdots]\langle T(J_1' J_1, t)_{KQ}^{\dagger}\rangle \tag{5.3.4a}$$

$$= [\cdots]\exp\left[-i\omega_{1'1} t - (\gamma_{1'} + \gamma_1)t/2\right]\langle T(J_1' J_1)_{KQ}^{\dagger}\rangle \tag{5.3.4}$$

That is, the state multipoles $\langle T(J_1' J_1)_{KQ}^{\dagger}\rangle$ characterize the atomic states immediately after the excitation and contain all information on the excitation process. The exponential factor describes the time evolution of the excited states between excitation and decay. The remaining factors in Eq. (5.2.7), indicated by the brackets in Eq. (5.3.4), are related to the decay process at time t.

Let us now consider the factors in brackets in Eq. (5.3.4) in more detail. Here the reduced matrix elements contain all information on the dynamics of the decay process, and the $3j$ and $6j$ symbols are geometrical factors depending on the angular momentum quantum numbers. The angular dependence of the emitted radiation is given explicitly by the elements of the rotation matrix. Any element $D_{qO}^{(K)}$ is related to the corresponding multipoles with the same K and Q. Thus *any state multipole with rank K and component Q present gives rise to a characteristic angular dependence of the emitted light*. As a consequence, by determining the elements $\dot{\rho}_{\lambda'\lambda}$ as a function of θ and φ information can be obtained on the state multipoles $\langle T(J_1'J_1)_{KO}^{\dagger}\rangle$. We will discuss this topic in more detail in Chapter 6.

Multipoles with the same K and Q but different J_1' and J_1 are related to the same element $D_{qO}^{(K)}$ and thus cannot be determined separately by measuring the angular distribution and polarization of the radiation. However, these multipoles are combined with different exponential factors and can therefore be determined separately by analyzing the time modulation of $\dot{\rho}_{\lambda'\lambda}$. Experiments which use this technique have been performed, for example, in the case of beam–foil-excited hydrogen atoms (see, for example, Burns and Hancock, 1971).

Of particular importance are the consequences of angular momentum conservation in the radiative decay expressed explicitly by the $3j$ and $6j$ symbols in Eqs. (5.2.6) and (5.2.7). We note that these symbols vanish for $K > 2$. This is a consequence of observation of dipole radiation where the total angular momentum carried away by the photons is $j = 1$. In general, if radiation with multipolarity j is detected then state multipoles with rank $K \leq 2j$ will contribute to $\rho_{\lambda'\lambda}$.

Thus, although all state multipoles satisfying $|J_1' - J_1| \leq K \leq J_1' + J_1$ are required for a complete description of the excited state density matrix,

▶ the elements $\rho_{\lambda'\lambda}$ depend only on the tensors with $K = 0$, $K = 1$, $K = 2$, and only these can be determined from an observation of dipole radiation.

The determination of the higher tensors with $K > 2$ requires, for example, observation of the emitted dipole radiation in the presence of external fields mixing tensors of rank $K \leq 2$ with tensors of rank $K > 2$ (see, for example, Section 6.3). Information on the higher multipoles can also be obtained by scattering electrons from laser-excited atoms [further details on this technique are given in the review by Hertel and Stoll (1978)].

In the special case of $J_1 = 0$ all atoms land in the same final state, and using Eq. (C11)

$$\begin{Bmatrix} 1 & 1 & K \\ J_1 & J_1' & 0 \end{Bmatrix} = \tfrac{1}{3}\delta_{J_11}\delta_{J_1'1}$$

In general the $6j$ symbol is smaller than $1/3$ if $J_2 \neq 0$. If J_1 is sharp then all elements $\rho_{\lambda'\lambda}$ and hence the values of the Stokes parameters and the degree of polarization P are reduced by a factor

$$3 \begin{Bmatrix} 1 & 1 & K \\ J_1 & J_1 & J_2 \end{Bmatrix}$$

compared to $J_0 = 0$. In fact, if $J_2 \neq 0$ the atoms land in states with different M_2 which are not detected in the experiment under discussion. As discussed in Section 3.3, the detected light is necessarily depolarized in the sense that $P < 1$. We can therefore interpret the $6j$ symbol as a *depolarization factor* which describes the depolarization of the radiation caused by the nonobservation of the final atomic states.

5.3.2. Manifestations of Coherence. Quantum Beats

Equation (5.2.7) shows that the angular distribution and polarization of the emitted photons are time modulated because of the presence of time-dependent factors with $J'_1 \neq J_1$. These factors give the time evolution of the state multipoles $\langle T(J'_1 J_1)^\dagger_{KQ} \rangle$ as expressed by Eq. (5.5.3). If the states with different J_1 have been incoherently excited then the density matrix $\rho(0)$ is diagonal in J_1 and, according to the discussion in Section 4.3.1, only state multipoles with $J'_1 = J_1$ can contribute to Eq. (5.2.7), all terms with $J'_1 \neq J_1$ vanish and no quantum beats occur. Thus *the observation of quantum beats can be regarded as a manifestation of the coherent excitation of states with different J_1 (different energy).*

Coherent excitation of states with different energies is only possible if the excitation process satisfies certain conditions. In the case of an isotropic excitation process (Section 4.5.2) all multipoles with $K \neq 0$ vanish. From Eqs. (5.2.6) and (5.2.7) and from the properties of the rotation matrix $D^{(K)}_{qQ}$ it follows that the photons are emitted isotropically and are unpolarized. The $6j$ symbols vanish unless $J'_1 = J_1$, in which case no quantum beats occur. Thus *anisotropic excitation is an essential requirement for the observation of any beat signal.* Furthermore, consider for example excitation of states from a ground state with sharp energy by photon absorption. Because of energy conservation coherent excitation requires that the exciting light contain a range of frequencies $\Delta\omega$ which is sufficiently broad to cover the difference $(1/\hbar)(E_{1'} - E_1) = \omega_{1'1}$ of levels with different energy (Section 3.1). Light pulses with a finite width $\Delta\omega$ can be represented by a coherent superposition of plane waves with different frequency, and *this coherence is transferred to the atoms* and is responsible for the quantum beats (see Section 2.3). Alternatively, coherent excitation can be considered a result of the requirement that the excitation time Δt is much shorter than any "characteristic" time of the excited states. Excitation times and energy spread of the exciting particles (electrons, photons) are related by the time–energy uncertainty relation $\Delta t \sim 1/\Delta\omega$. The "characteristic" time is the time interval $\Delta t_c \sim 1/\omega_{1'1}$ with $\omega_{1'1} = (1/\hbar)(E_{1'} - E_1)$ determining the largest energy difference of the excited states. From the condition that $\Delta t \ll \Delta t_c$ it follows once more that the condition for coherent excitation is that the energy uncertainty of the exciting particles must cover the energy difference of several excited states.

Let us now consider the observational effects of coherence between states with the same J_1 but different M_1. If these states can be considered to be degenerate the coherence does not produce quantum beat effects. If the magnetic substates have been incoherently excited $\rho(0)$ is diagonal in M_1. As discussed in Section 4.5, the source is then axially symmetric with respect to the Z axis of the collision system and all tensors with $Q \neq 0$ vanish. The angular dependence of the elements $\rho_{\lambda'\lambda}$ is then determined by the elements of the rotation matrix with $Q = 0$ which are represented by the "small" d functions:

$$D(0\theta\varphi)_{q0}^{(K)} = d(\theta)_{q0}^{(K)}$$

(with $q = \lambda' - \lambda$). Consequently, the elements $\rho_{\lambda'\lambda}$ depend only on the polar angle θ of $\mathbf{n}$ but not on the azimuthal angle φ: hence the emitted light is axial symmetric about the Z axis of the collision system.

If different magnetic substrates have been excited coherently then multipoles with $Q \neq 0$ will be different from zero and the elements $\rho_{\lambda'\lambda}$ will depend on the azimuthal angle φ. Thus *coherence between states with the same J_1 but different M_1 manifests itself in a change of the angular dependence of the emitted radiation*.

Consider now the case of an axially symmetric atomic source which is aligned but not oriented. [This can be obtained, for example, by exciting the atoms by electron impact and not observing the scattered electrons (Section 4.6.3) or by absorption of unpolarized or linearly polarized photons (Section 4.5.3).] Because of the relation

$$\begin{pmatrix} 1 & 1 & K \\ -\lambda' & \lambda & q \end{pmatrix} = (-1)^K \begin{pmatrix} 1 & 1 & K \\ \lambda' & -\lambda & -q \end{pmatrix}$$

it follows that $\rho_{11} = \rho_{-1-1}$ and thus $\eta_2 = 0$. Hence, in this case, the emitted light contains photons in both helicity states with equal intensity, and the degree of circular polarization vanishes.

Finally, let us assume that the geometry of the excitation process contains a plane of symmetry. This applies for the case discussed in Section 4.6. For sharp J_1 the elements $\rho_{\lambda'\lambda}$ depend on four parameters (besides the monopole), one component of the orientation vector and three components of the alignment tensor. This situation occurs also in the case of beam–foil-excited atoms when the foil axis is tilted with respect to the incoming beam axis.

5.4. Perturbed Angular Distribution and Polarization

5.4.1. General Theory

In this section we will consider the case in which the excited atoms are perturbed by an external or internal field. The theory of perturbed angular distributions has been developed in nuclear physics (see, for example, Steffen and Alder, 1975). Here, we will begin by discussing the basic principles of the theory and in the following section and in Chapter 6 we will give some examples of its use.

In what follows it will always be assumed that the perturbation is weak and of little relevance to excitation and decay processes but is sufficiently strong to change the state of the atoms between excitation and decay (assuming sharply defined excitation and decay times). In this case the theory developed in Section 4.7 applies.

If the perturbation can be neglected during the excitation process the excited atomic states at time $t = 0$ can be characterized in terms of state multipoles $\langle T(J_1'J_1)_{KQ}^\dagger\rangle$. The subsequent time evolution of the multipoles is now governed by a Hamiltonian $H = H_0 + H'$ (where H' is the perturbation term), and the evolution operator is $U = \exp[-(i/\hbar)Ht - \Gamma t/2]$. Consequently, $U(t)$ replaces $U(t)_0$ in the equations of Section 5.3. The excited atoms at time t are characterized by the state multipoles:

$$\langle T(J_1'J_1t)_{KQ}^\dagger\rangle = \sum_{\substack{K'Q' \\ j_1'j_1}} \langle T(j_1'j_1)_{K'Q'}^\dagger\rangle G(j_1'j_1J_1'J_1, t)_{K'K}^{Q'Q} \qquad (5.4.1)$$

as in the general equation (4.7.5) where the radiative decay term has now been included. Substitution of Eq. (5.4.1) for $\langle T(J_1'J_1)_{KQ}^\dagger\rangle$ into Eq. (5.3.4a) yields

$$\dot\rho(\mathbf{n}, t)_{\lambda'\lambda} = C(\omega) \sum_{\substack{J_1'J_1j_1'j_1 \\ KK'Q'Q}} \mathrm{tr}\,[r_{-\lambda'}T(J_1'J_1)_{Kq}r_{-\lambda}^\dagger]$$

$$\times D(0\theta\varphi)_{qQ}^{(K)}G(j_1'j_1J_1'J_1, t)_{K'K}^{Q'Q}\langle T(j_1'j_1)_{K'Q'}^\dagger\rangle \qquad (5.4.2)$$

By comparing with Eq. (5.3.4) it can be seen that the time-dependent exponential factor in Eq. (5.3.4) is replaced by the general perturbation coefficient describing the time evolution.

Various external and internal perturbations may influence the time evolution, and by observing experimentally the way in which these perturbations affect the angular distribution and polarization of the emitted light it is possible to extract information on various properties of the excited states. We will now illustrate this with some examples.

5.4.2. Quantum Beats Produced by "Symmetry Breaking"

In Sections 5.1–5.3 it was assumed that atomic states with different angular momenta J_1 had been coherently excited at time $t = 0$. This coherency leads to a time modulation of the exponential decay of the excited states. No quantum beats will occur if the excitation is incoherent.

This conclusion may not be correct if the atomic lifetime is sufficiently long and if the excited states are perturbed by external or internal fields during the time between excitation and decay. These perturbations will disturb the excited states and lead to time modulation of the orientation and alignment parameters and, hence, of the angular distribution and polarization of the emitted light, even if there is no coherence between the initially excited states. In this case *the perturbation subsequent to the excitation forms the basis for the quantum beats.*

A clear discussion of the underlying principles has been given, for example, by Series and Dodd (1978) and Andrä (1979). *The essence of the method is a sudden change in the Hamiltonian describing the excited atoms.* If at times $t < 0$ the atoms are in eigenstates $|\varphi_0\rangle$ of H_0 and a sudden change in the Hamiltonian from H_0 to H is introduced at time $t = 0$, then for $t > 0$ the time evolution is governed by H. Any eigenstate $|\varphi_0\rangle$ will evolve into a coherent superposition of eigenstates of H and this coherence gives rise to quantum beats.

This general principle applies, for example, to situations where a beam of free atoms enters an external field with a sudden onset for the passing beam. Another example is provided by the case described in Sections 3.5 and 4.6. An atomic ensemble is "instantaneously" excited at $t = 0$ into eigenstates $|LMS_1M_{S_1}\rangle$ of a Hamiltonian H_0 with all explicit spin couplings neglected. The subsequent time evolution for $t > 0$ is then governed by the full Hamiltonian H of the free atoms including fine-structure (and possibly hyperfine-structure) effects. A similar situation is encountered in beam–foil excitation. Here the atoms are assumed to be excited into uncoupled states $|LMS_1M_{S_1}\rangle$ during the short time interval in which the atoms pass through the foil. After emerging from the foil the atoms evolve under the influence of the full Hamiltonian and explicit spin-dependent coupling terms must be taken into account.

As an illustration of this let us consider the situation in which an atomic ensemble is assumed to be excited "instantaneously" at time $t = 0$ with the spins unaffected and evolves for $t > 0$ under the influence of the fine-structure interaction.

We obtain the elements $\dot{\rho}(\mathbf{n}, t)_{\lambda'\lambda}$ of the density matrix of the photons, emitted at time t, by inserting the relevant perturbation coefficient in Eq. (5.4.2). The perturbation coefficient for fine-structure interaction is given by

Eq. (4.7.17). Taking into account the radiative decay we substitute

$$U(t) = \exp\left(-iHt/\hbar - \Gamma t/2\right)$$

and obtain a perturbation coefficient $G(L, t)_K \exp(-\gamma t)$, where $G(L, t)_K$ is given by Eq. (4.7.17). In doing this it is assumed that all fine-structure states belonging to the same level (LS_1) have the same decay constant γ. We then obtain from Eq. (5.4.2)

$$\dot{\rho}(\mathbf{n}, t)_{\lambda'\lambda} = C(\omega) \sum_{KQq} \mathrm{tr}\left[r_{-\lambda'} T(L)_{Kq} r^\dagger_{-\lambda}\right] D(\theta\varphi)^{(K)}_{qQ} G(L, t)_K$$

$$\times \exp(-\gamma t) \langle T(L)^\dagger_{KQ} \rangle \tag{5.4.3a}$$

In particular, for the intensity $I(\mathbf{n}, t)$ of the light emitted at time t in the direction $\mathbf{n}$ is given by

$$I(\mathbf{n}, t) = \dot{\rho}(\mathbf{n}, t)_{11} + \dot{\rho}(\mathbf{n}, t)_{-1-1}$$

$$= C(\omega) \sum_{L_2 KQ} |\langle L_2 \| \mathbf{r} \| L \rangle|^2 (-1)^{L+L_2} C(K)$$

$$\times \begin{Bmatrix} 1 & 1 & K \\ L & L & L_2 \end{Bmatrix} Y(\theta\varphi)_{KQ} \frac{\exp(-\gamma t)}{2S_1 + 1}$$

$$\times \left[\sum_{J'_1 J_1} (2J'_1 + 1)(2J_1 + 1) \begin{Bmatrix} L & J'_1 & S_1 \\ J_1 & L & K \end{Bmatrix}^2 \cos \omega_{1'1} t \right] \langle T(L)^\dagger_{KQ} \rangle \tag{5.4.3b}$$

where $C(K)$ denotes the numerical factors:

$$C(0) = -2(4\pi/3)^{1/2}, \qquad C(1) = 0, \qquad C(2) = -(8\pi/15)^{1/2} \tag{5.4.4}$$

Equation (5.4.3) shows that direct measurements of fine-structure splittings are possible merely by observing the emitted light as a function of time (see Section 2.3).

For $K = 0$ the $6j$ symbols within the brackets in Eq. (5.4.3) reduce to the expression

$$\begin{Bmatrix} L & J'_1 & S_1 \\ J_1 & L & 0 \end{Bmatrix} = \frac{(-1)^{S_1 + L + J_1}}{[(2L + 1)(2J_1 + 1)]^{1/2}} \delta_{J'_1 J_1} \tag{5.4.5}$$

and all interference terms with $J'_1 \neq J_1$ vanish. Thus *orientation and/or alignment of the atomic source is essential for the observation of fine-structure quantum beats.*

Since the fine-structure interaction is isotropic it does not relate multipoles with different K and Q as shown by Eq. (4.7.18). Thus the initial symmetry is maintained for all times $t > 0$. If, for example, the initially excited states are axially symmetric with respect to some axis the light emission will be axially symmetric with respect to the same axis irrespective of the perturbation.

Similar results hold for the hyperfine interaction. The joint effect of fine and hyperfine interaction can be taken into account by substituting the relevant perturbation coefficient (4.7.22) for $G(L_1 t)_K$ in Eq. (5.4.3a).

In the treatment given here we have considered separately the effects due to the coherent excitation of nondegenerate states (Sections 5.1–5.3) and the effects caused by the fine-structure interaction. In general, both effects will overlap and the corresponding quantum beats will be superposed on each other. Such situations have been analyzed experimentally, for example, by Burns and Hancock (1971) (see also the review of Macek and Burns, 1976, and the references therein).

In conclusion, it has been seen that *the observation of quantum beats requires* (i) *a well-defined excitation time (pulsed excitation with pulses shorter than any characteristic atomic time), and* (ii) *time-resolved detection of the emitted light* together with an observation time resolution $t_R \gtrsim 1/\omega_{1'1}$. The lack of appropriate equipment with sufficient time resolution initially limited quantum beat experiments to Zeeman effect studies where the level splittings can be adjusted by the magnetic field to a few megahertz. A wider use of quantum beats has only become possible after the advent of lasers with pulse lengths of nanoseconds and in particular of beam–foil excitation, which allows excitation times of the order of 10^{-14} sec.

Experimental details, results, and further discussions of the method may be found in recently published reviews and books. We particularly refer to Corney (1977) and to the various chapters in Hanle and Kleinpoppen (1978 and 1979).

5.5. Time Integration over Quantum Beats

5.5.1. Steady State Excitation

Let us now return to Eq. (5.2.6). This expression describes, for example, the following situation. An atomic ensemble has been excited at time $t = 0$ and the radiation, emitted in the subsequent decay, is observed in the direction **n**. The photon detector may have a resolution time t_R so that all photons emitted in the time interval $0 \cdots t_R$ are taken into account. The corresponding density matrix elements are then given by Eq. (5.2.6) with the substitution $t \rightarrow t_R$.

The periods $\omega_{1'1}$ and the mean lifetime $\tau = \gamma^{-1}$ are much shorter than commonly employed resolution times so that the factor $\exp(-\gamma t_R)$ is effectively zero. Substituting this into Eq. (5.2.6) gives

$$\rho(\mathbf{n})_{\lambda'\lambda} = C(\omega) \sum_{J'J_1J_2} \langle J_2\|r\|J'_1 \rangle \langle J_2\|r\|J_1 \rangle^* (-1)^{J_1-J_2+\lambda}$$

$$\times \sum_{KQq} (2K+1)^{1/2} \begin{pmatrix} 1 & 1 & K \\ -\lambda' & \lambda & q \end{pmatrix} \begin{Bmatrix} 1 & 1 & K \\ J'_1 & J_1 & J_2 \end{Bmatrix}$$

$$\times D(\mathbf{n})_{qQ}^{(K)} \frac{1}{i\omega_{1'1}+\gamma} \langle T(J'_1J_1)_{KQ}^{\dagger} \rangle \tag{5.5.1}$$

where we have put $\gamma_i = \gamma$ for all i and where $\rho(\mathbf{n})_{\lambda'\lambda}$ denotes the time-integrated density matrix elements for $t_R \gg \tau$.

Equation (5.5.1) applies in particular to *steady state excitation* (starting at time $t = 0$), for example, by wave packets of light emitted by a resonance lamp, or by a current of electrons. In this and the following chapter we will use the *pulse approximation*. That is, we will always approximate the incoming flux by a succession of random pulses at random times. This causes the fluctuations to be smoothed out. For a discussion of this point we refer to the article by Series and Dodd (1978) (see also Chapter 7).

Since in the case of steady state excitation the time at which the emitted photons are observed is no longer well defined with respect to the time at which excitation took place, it is necessary to integrate over all observation times and Eq. (5.5.1) applies. The interaction between the atoms and a sequence of randomly phased pulses is an incoherent process. Each pulse gives rise to a radiation described by Eq. (5.5.1). The total density matrix is then given by $N\rho(\mathbf{n})_{\lambda'\lambda}$, where N is the number of pulses per second.

It is important to realize that the terms $\sim \langle T(J'_1J_1)_{KQ}^{\dagger} \rangle$, characterizing the coherence between the initial states with different J_1, vanish if the levels do not overlap, that is, if $\omega_{1'1}$ is much larger than the linewidth: $\omega_{1'1} \gg \gamma$. In this case

$$\left| \frac{1}{i\omega_{1'1}+\gamma} \right| \ll \frac{1}{\gamma} \tag{5.5.2}$$

and the dominant contributions to $\rho_{\lambda'\lambda}$ come from the incoherent terms with $J'_1 = J_1$. Alternatively, if the mean life time τ is much larger than the times $\omega_{1'1}^{-1}$ many oscillations will take place during the lifetime of the atoms, and will practically cancel each other in all time-dependent expressions. *In this case the initial coherence between the atomic states with different energies has no observational effect.*

We may summarize our results as follows. If a number of states $|J_1M_1\rangle$ with different J_1 and energy are coherently excited (which requires a sufficiently short excitation time as discussed in Section 5.4) then this coherence leads to quantum beats. (Compare, for example, Figure 2.1 with

Figure 3.1 and see the discussion in Section 3.4.2.) The corresponding interference terms can be directly observed in experiments with sufficiently high time resolution and well-defined excitation time. *If, however, time-integrated quantities are observed (as is always the case in steady state excitation) the coherence between states with different energy is retained when the energy separation of the excited states is small compared to their width, the coherence being destroyed when the levels are well separated.*

5.5.2. Depolarization Effects Caused by Fine and Hyperfine Interaction

Let us now consider the case in which the assumptions of Section 5.4.2 apply but where the resolution time is not high enough to observe quantum beats. Equation (5.4.3a) must then be integrated over a time interval $0 \cdots t_R$. Assuming that t_R *is much larger than the mean lifetime of the excited states* we can extend the upper limit to infinity with negligible error. The integral over the perturbation coefficients $G(Lt)_K \exp(-\gamma t)$ gives

$$G(L)_K \equiv \int_0^\infty dt\, G(L, t)_K \exp(-\gamma t)$$

$$= \frac{1}{2S_1 + 1} \sum_{J'_1 J_1} (2J'_1 + 1)(2J_1 + 1) \begin{Bmatrix} L & J'_1 & S_1 \\ J_1 & L & K \end{Bmatrix}^2 \frac{\gamma}{\gamma^2 + \omega^2_{1'1}} \quad (5.5.3)$$

The time-integrated density matrix elements are then given by

$$\rho(\mathbf{n})_{\lambda'\lambda} = C(\omega) \sum_{KQq} \mathrm{tr}\,[r_{-\lambda'} T(L)_{Kq} r^\dagger_{-\lambda}] D(\mathbf{n})^{(K)}_{qQ} G(L)_K \langle T(L)^\dagger_{KQ} \rangle \quad (5.5.4)$$

For singlet–singlet transitions we have $G(L)_K = 1/\gamma$. For $S_1 \neq 0$ it can be shown, using the properties of the $6j$ symbols, that $G(L)_K < 1/\gamma$ for $K \neq 0$. Because of this relation the quantities $G(L)_K \langle T(L)^\dagger_{KQ} \rangle$ are smaller than the corresponding parameters $(1/\gamma)\langle T(L)^\dagger_{KQ} \rangle$ in the singlet case: *the coupling of the orbital system to the unpolarized spins results in a loss of orientation and alignment.* In addition to this there is a second depolarization effect if $L_2 \neq 0$ (see the discussion in Section 5.3.1).

In a similar way the influence of hyperfine interaction on the emitted radiation may be discussed simply by substituting the relevant perturbation coefficient (4.7.21) or (4.7.22) for $G(L, t)_K$ in Eq. (5.4.3a).

It is instructive to consider the extreme cases in which the linewidth is either much larger or much smaller than the fine-structure splitting $\omega_{1'1}$. In the first case we have

$$\frac{\gamma}{\gamma^2 + \omega^2_{1'1}} \approx \frac{1}{\gamma} \quad (5.5.5)$$

for *all* terms in the equations for the Stokes parameters. From Eq. (5.5.3) and the orthogonality conditions of the $6j$ symbols [Eq. (C10)] we obtain for all K

$$G(L)_K = \frac{1}{\gamma} \frac{1}{2S_1 + 1} \sum_{J_1' J_1} (2J_1' + 1) \begin{Bmatrix} L & J_1' & S_1 \\ J_1 & L & K \end{Bmatrix}^2 (2J_1 + 1)$$

$$= \frac{1}{\gamma} \frac{1}{(2S_1 + 1)(2L + 1)} \sum_{J_1'} (2J_1' + 1)$$

which gives

$$G(L)_K = 1/\gamma \qquad (5.5.6)$$

which are the same values as in the spinless case.

The result is readily understood since in the case under discussion the mean lifetime of the excited states $\tau \sim \gamma^{-1}$ is small compared to the precession times $\omega_{1'1}^{-1}$ associated with the spin–orbit coupling. That is, the atoms emit the photons before the precessional motion can be set up. Fine-structure interaction can then be neglected and the expressions for the Stokes parameters are the same as in the case of spinless atoms.

Consider now the case $\gamma \ll \omega_{1'1}$ or, alternatively, $\tau \gg \omega_{1'1}^{-1}$. Here,

$$\frac{\gamma}{\gamma^2 + \omega_{1'1}^2} \begin{cases} = \dfrac{1}{\gamma} & \text{for } J_1' = J_1 \\[2mm] \approx \dfrac{1}{\gamma} \cdot \dfrac{\gamma^2}{\omega_{1'1}^2} \ll \dfrac{1}{\gamma} & \text{for } J_1' \neq J_1 \end{cases} \qquad (5.5.7)$$

Thus, as discussed in Section 5.5.1, the main contributions to the Stokes parameters stem from the terms with $J_1' = J_1$, and the interference terms with $J_1' \neq J_1$ can be neglected. In this case it follows from Eqs. (5.5.3) and (5.5.7) that

$$G(L)_K = \frac{1}{\gamma} \frac{1}{2S_1 + 1} \sum_{J_1} (2J_1 + 1)^2 \begin{Bmatrix} L & J_1 & S_1 \\ J_1 & L & K \end{Bmatrix}^2 \qquad (5.5.8)$$

Since $G(L)_K < 1/\gamma$ for $K \neq 0$ anisotropy and polarization of the emitted radiation are reduced.

This result can be understood by realizing that in the case under consideration many precessions take place during the atomic lifetime. Since we are interested in quantities averaged over a time interval $0 \cdots t_R$ with $t_R \gg \tau$ all interference terms practically cancel each other and only the time-independent terms with $J_1' = J_1$ will survive (compare this with the discussion in Section 4.7.3).

In conclusion, if the fine-structure separation is comparable with the linewidths expression (5.5.3) has to be used. If the linewidth is much larger than the energy separation $\omega_{1'1}$ fine-structure effects can be neglected. If $E_{1'} - E_1 \gg \hbar\gamma \gg E_{J_1 F_1'} - E_{J_1 F_1}$ fine-structure effects must be taken into account but hyperfine interaction can be neglected. The relevant factors $G(L)_K$ are given by Eq. (5.5.8). If γ is small compared to the hyperfine splitting $(E_{J_1 F_1'} - E_{J_1 F_1})$ then hyperfine interaction must be taken into account. The corresponding expressions for the Stokes parameters are obtained simply by substituting the relevant perturbation coefficient (4.7.22) for $G(Lt)_K$ into all the above formulas.

6

Some Applications

6.1. Theory of Electron–Photon Angular Correlations in Atomic Physics

6.1.1. Singlet–Singlet Transitions

The fundamental equations (5.2.6), (5.2.7), and (5.4.2) can be applied to a variety of experimental situations and in this chapter we will give some instructive examples of their utility. As the first example, we will show how *information on the excitation process* can be obtained from a determination of the Stokes parameters. In particular, we will consider the case of excitation of atoms by electron impact under the conditions described in Sections 3.5 and 4.6.

All information on the collision process is contained in the relevant reduced density matrix $\rho(0)$ describing the excited atomic states immediately after the excitation. Its complete determination requires the measurement of all independent state multipoles of rank $K \leq 2L$ and component Q satisfying $-K \leq Q \leq K$. If the scattered electrons are not observed only tensors with $Q = 0$ can be nonzero. It is possible to obtain more information on the excitation process if scattered electrons and emitted photons are detected in coincidence. In this case the observation is restricted to light emitted by an atomic subensemble only, namely, those atoms excited by the detected electrons. It was shown in Section 4.6 that the atomic subensemble of interest is characterized by a monopole, one component of the orientation vector, three components of the alignment tensor, and all independent tensors with higher rank $2 < K \leq 2L$. However, the multipoles with $K > 2$ cannot be obtained from an observation of dipole radiation unless perturbations are present mixing tensors with different rank (see Section 5.3). The electron–photon coincidence experiments to be discussed here allow the determination of four parameters in addition to the differential cross section σ. An experimental determination of these parameters and comparison with theoretical results provides more sensitive tests for theoretical predictions than the more traditional experiments in which only σ is determined. In this section we will consider the excitation of singlet states from an atomic ground state for which $S_0 = 0$ and $L_0 = 0$. Suppose that light emitted in a transition $L \to L_2$ is observed ($S_2 = 0$) and that the resolution time t_R of the photon detector is much larger than the mean lifetime of the excited atoms. Assuming that no perturbations are present the polarization density matrix of the photons emitted in the time interval $0 \cdots t_R$ is given by Eq. (5.5.1) with L replacing J_1 and J_1'.

Explicit expressions for the Stokes parameters can be derived from the polarization density matrix. The emitted photons are observed in the direction of the unit vector $\mathbf{n}$ with polar angles θ and φ in the collision system. The Stokes parameters are most conveniently discussed in a coordinate system where $\mathbf{n}$ is the quantization axis. The polarization vector of the emitted light is restricted to the plane perpendicular to the direction of propagation $\mathbf{n}$. Two orthogonal unit vectors $\mathbf{e}_1$ and $\mathbf{e}_2$ can be chosen to span this plane (see Section 1.2). $\mathbf{e}_1$ is chosen to lie in the plane formed by $\mathbf{n}$ and the Z axis and to point in the direction of increasing θ. $\mathbf{e}_2$ is then chosen to be perpendicular to both $\mathbf{e}_1$ and $\mathbf{n}$ and to point in the direction of increasing φ. The vector $\mathbf{e}_1$ then has polar angles $(\theta + 90°, \varphi)$ in the collision system and the vector $\mathbf{e}_2$ has the polar angles $(90°, \varphi + 90°)$. $\mathbf{e}_1$ therefore has the same azimuth angle as $\mathbf{n}$ and $\mathbf{e}_2$ lies in the X–Y plane at an angle φ to the Y axis (see Figure 6.1). In this "detector" system the Stokes parameter η_3 is the degree of linear polarization in the direction $\mathbf{e}_1$ and η_1 the degree of linear

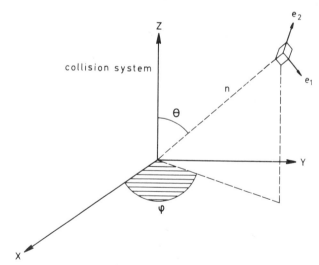

Figure 6.1. Coordinate systems used in the description of coincidence experiments.

polarization at angles $\pm 45°$ to $\mathbf{e}_1$. The Stokes parameters can be calculated from Eqs. (5.4.3a), (5.2.4) (with $J_1' = J_1 = L$), (1.2.23), and substituting explicit expressions for the rotation matrix elements. This gives

$$I = \frac{C(\omega)}{\gamma}|\langle L_2\|\mathbf{r}\|L\rangle|^2(-1)^{L+L_2}\left[\frac{2(-1)^{L+L_2}}{3(2L+1)^{1/2}}\langle T(L)_{00}\rangle\right.$$ (6.1.1a)

$$-\begin{Bmatrix}1 & 1 & 2\\ L & L & L_2\end{Bmatrix}\left(\langle T(L)_{22}^\dagger\rangle \sin^2\theta\cos 2\varphi - \langle T(L)_{21}^\dagger\rangle \sin 2\theta\cos\varphi\right.$$

$$\left.\left.+\frac{\langle T(L)_{20}^\dagger\rangle(3\cos^2\theta - 1)}{6^{1/2}}\right)\right]$$

$$I\eta_3 = \frac{C(\omega)}{\gamma}|\langle L_2\|\mathbf{r}\|L\rangle|^2(-1)^{L+L_2}\begin{Bmatrix}1 & 1 & 2\\ L & L & L_2\end{Bmatrix}\left[\langle T(L)_{22}^\dagger\rangle(1+\cos^2\theta)\cos 2\varphi\right.$$

$$\left.+\langle T(L)_{21}^\dagger\rangle \sin 2\theta\cos\varphi + \left(\frac{3}{2}\right)^{1/2}\langle T(L)_{20}^\dagger\rangle \sin^2\theta\right]$$ (6.1.1b)

$$I\eta_1 = -\frac{C(\omega)}{\gamma}|\langle L_2\|\mathbf{r}\|L\rangle|^2(-1)^{L+L_2}\begin{Bmatrix}1 & 1 & 2\\ L & L & L_2\end{Bmatrix}[\langle T(L)_{22}^\dagger\rangle 2\cos\theta\sin 2\varphi$$

$$+\langle T(L)_{21}^\dagger\rangle 2\sin\theta\sin\varphi]$$ (6.1.1c)

$$I\eta_2 = -\frac{C(\omega)}{\gamma}|\langle L_2\|\mathbf{r}\|L\rangle|^2(-1)^{L+L_2}\begin{Bmatrix}1 & 1 & 1\\ L & L & L_2\end{Bmatrix}2i\langle T(L)_{11}^\dagger\rangle \sin\theta\sin\varphi$$

(6.1.1d)

Note that the tensors $\langle T(L)_{2Q}^\dagger\rangle$ and $i\langle T(L)_{11}^\dagger\rangle$ are real quantities (see Section 4.6). We have used the approximation 5.5.

The monopole $\langle T(L)_{00}\rangle$ can be obtained from a measurement of the differential cross section. The equations (6.1.1) afford then several possibilities for determining the state multipoles with $K > 0$. For example, I can be measured for three different pairs of angles θ, φ (this gives three equations which allow the extraction of the three independent components of the alignment tensor) and $I\eta_2$ obtained at one set of angles θ, φ (which gives the orientation vector). Alternatively, all four Stokes parameters can be measured in the same direction θ, φ, and by inserting the obtained values into Eqs. (6.1.1) four equations are obtained from which the orientation and alignment parameters can be extracted. Using these methods these excitation parameters have been experimentally determined for several atoms and compared with theoretical predictions [for further details see the review by Blum and Kleinpoppen (1979)].

In general, all five multipoles contributing to Eqs. (6.1.1) are independent. The four Stokes parameters and the differential cross section σ are therefore also independent quantities. In particular, the angular distribution I contains information on the atomic source which cannot be obtained from a determination of the other parameters.

The first experiment of this kind was carried out by Eminyan, McAdam, Slevin and Kleinpoppen (1974) for ^1P excitation of helium. In this case ($L = 1$) the excited atoms are completely characterized in terms of three parameters only (see Section 3.5.2). Let us discuss this case in more detail. The Stokes parameters are not independent and only three independent measurements are required for a complete determination of the atomic density matrix, for example σ, I, and $I\eta_2$. Expressing the multipole parameters in terms of the parameters σ, λ, χ introduced in Section 3.5.2 we obtain from Eqs. (6.1.1) with $L = 1$, $L_2 = 0$

$$I = \frac{C(\omega)}{\gamma} |\langle 0\|\mathbf{r}\|1\rangle|^2 \frac{\sigma}{3} \left[\frac{1-\lambda}{2} (1 - \sin^2 \theta \cos 2\varphi + \cos^2 \theta) + \lambda \sin^2 \theta \right.$$
$$\left. + [\lambda(1-\lambda)]^{1/2} \cos \chi \sin 2\theta \cos \varphi \right] \tag{6.1.2a}$$

$$I\eta_2 = + \frac{C(\omega)}{\gamma} |\langle 0\|\mathbf{r}\|1\rangle|^2 \frac{2\sigma}{3} [\lambda(1-\lambda)]^{1/2} \sin \chi \sin \theta \sin \varphi \tag{6.1.2b}$$

where we inserted explicit values for the $6j$ symbols. It follows that in this case, a complete determination of the scattering amplitudes is possible. For this reason the case $L = 1$ and $S_1 = 0$ is of particular interest.

From Eq. (6.1.2b) it follows that the degree of circular polarization is determined by the phase χ and, correspondingly, a measurement of $I\eta_2$ directly determines χ. If the photons are detected in the Y direction ($\theta = \varphi = 90°$) the degree of circular polarization is given by dividing Eqs. (6.1.2a) and (6.1.2b):

$$\eta_2 = +2[\lambda(1-\lambda)]^{1/2} \sin \chi \tag{6.1.3}$$

By expressing the scattering amplitudes in Eq. (4.6.1) in terms of σ, λ, χ according to Section 3.5.2 it can be shown with the help of Eq. (4.6.5a) that

$$\eta_2 = \langle L_y \rangle \tag{6.1.4}$$

Thus η_2 is a direct measure of the degree of orientation, or, of the net amount of the angular momentum transferred to the atoms during the excitation process.

When the photons are detected in the scattering plane φ is zero and

$$\eta_3 = +1 \tag{6.1.5}$$

This can be shown by first specializing Eq. (6.1.1b) to the case $L = 1, L_2 = 0$:

$$I\eta_3 = -\frac{C(\omega)}{\gamma}|\langle 0\|\mathbf{r}\|1\rangle|^2 \frac{\sigma}{3}\left\{\frac{1-\lambda}{2}[(-1-\cos^2\theta)\cos 2\varphi + \sin^2\theta]\right.$$

$$\left. -\lambda\sin^2\theta - [\lambda(1-\lambda)]^{1/2}\cos\chi\sin 2\theta\cos\varphi\right\} \qquad (6.1.6)$$

Inserting $\varphi = 0$ and dividing by Eq. (6.1.2a) we obtain the relation (6.1.5). Thus the photons observed in the scattering plane are completely linearly polarized. The electric vector oscillates along the direction of $\mathbf{e}_1$.

As discussed in Section 3.5.2 the excited ^1P state is a completely coherent superposition of the magnetic substate. From this and the fact that $L_2 = 0$ it follows that the detected light is emitted in a transition between two pure atomic states. As a result, the light is necessarily completely polarized in the sense that $P = 1$.

6.1.2. Influence of Fine and Hyperfine Interaction on the Emitted Radiation

We will now discuss the excitation of atomic levels with orbital angular momentum L and spin $S_1 \neq 0$ by electron impact and the coincident detection of scattered electrons and emitted photons. We will assume that the assumptions discussed in Sections 3.5 and 4.7.2 apply.

Immediately after the excitation the state of the atomic subensemble of interest is characterized by the state multipoles (4.6.1). These are perturbed by the fine (and possibly hyperfine) interaction which in turn influences the emitted light.

Assuming that only *time-integrated quantities* are observed (with an upper limit $t_R \gg \tau$) the density matrix elements of the emitted radiation are given by Eq. (5.5.4) where now the state multipoles are given by Eq. (4.6.1). Recalling that $G(L)_K = 1/\gamma$ in the case of spinless atoms the Stokes parameters in the present case of interest can be obtained by substituting $G(L)_K\langle T(L)_{KQ}^\dagger\rangle$ for $1/\gamma\langle T(L)_{KQ}^\dagger\rangle$ in Eqs. (6.1.1). Thus, for example, the angular distribution becomes, with $G(L)_0 = 1/\gamma$,

$$I = \frac{C(\omega)}{\gamma}|\langle L_2\|\mathbf{r}\|L\rangle|^2\left\{\frac{2}{3(2L+1)^{1/2}}\langle T(L)_{00}^\dagger\rangle\right.$$

$$-(-1)^{L+L_2}\gamma G(L)_2\begin{Bmatrix}1 & 1 & 2\\ L & L & L_2\end{Bmatrix}[T(L)_{22}^\dagger]\sin^2\theta\cos 2\varphi$$

$$\left. -\langle T(L)_{21}^\dagger\rangle\sin 2\theta\cos\varphi + \frac{\langle T(L)_{20}^\dagger\rangle(3\cos^2\theta - 1)}{6^{1/2}}\right]\right\} \qquad (6.1.7)$$

In order to obtain the parameters $I\eta_3$ and $I\eta_1$ Eqs. (6.1.1b) and (6.1.1c) must be multiplied by a common factor $\gamma G(L)_2$, and $I\eta_2$ is obtained by multiplying Eq. (6.1.1d) by $\gamma G(L)_1$.

In these expressions the loss of orientation and alignment, caused by the coupling to the unobserved spin system, is described by the factors $G(L)_1$ and $G(L)_2$. Since $G(L)_K < 1/\gamma$ for $K \neq 0$ Eq. (6.1.7) shows that the angular distribution becomes more isotropic compared to the spinless case (the angular distribution is "smeared out" as a result of spin–orbit coupling). The values of the other Stokes parameters are reduced from their values for spinless atoms by a factor $\gamma G(L)_1$ and $\gamma G(L)_2$, respectively, which results in a depolarization of the emitted radiation.

The perturbation factors $G(L)_K$ are given by Eq. (5.5.3) if the fine-structure separation $\omega_{1'1}$ and linewidth are comparable, by Eq. (5.5.8) if the fine-structure levels do not overlap ($\gamma \ll \omega_{1'1}$), or by Eq. (5.5.6) if $\gamma \gg \omega_{1'1}$. In the latter case the fine-structure interaction does not affect the emitted light.

The effect of hyperfine interaction can be treated by the method outlined in Section 5.5.2.

6.2. Steady State Excitation

6.2.1. Polarization of Impact Radiation

The formulas presented in Section 6.1 describe the polarization properties of light detected in coincidence with the scattered electrons, that is, those photons which are only emitted by a particular subensemble of atoms. We will now consider the case where the scattered electrons are not observed. It will be assumed that the resolution of the photon detector is sufficient to restrict the observation to photons emitted in a transition between levels with fixed quantum numbers $LS_1 \rightarrow L_2S_2$. Since the electrons are not observed, however, all photons emitted in this transition must now be taken into account irrespective of the direction in which the electrons are scattered.

If we now consider excitation by a steady flux of incoming electrons the time at which the photons are emitted is no longer uniquely defined with respect to the excitation time. Thus the time-integrated form of the polarization density matrix must be used.

When the scattered electrons are unobserved the atomic system of interest is aligned but not oriented according to the discussions in Section 4.6.3. Consequently, the detected radiation depends on only two parameters, the monopole $\langle T(L)_{00} \rangle$, which is proportional to the total cross section Q, and the alignment parameter $\langle T(L)_{20}^{\dagger} \rangle$ given by Eq. (4.6.11). Thus the relevant Stokes parameters are readily obtained by putting all multipoles equal to zero in Eq. (6.1.7) and the equations for the other Stokes parameters except $\langle T(L)_{00}\rangle$ and $\langle T(L)_{20}^{\dagger}\rangle$. The degree of circular polarization and the parameter η_1 vanish and the nonzero Stokes parameters are found to be

$$I(\mathbf{n}) = C(\omega)|\langle L_2\|\mathbf{r}\|L\rangle|^2 \frac{2}{3(2L+1)^{1/2}}\langle T(L)_{00}\rangle\frac{1}{\gamma}$$

$$-C(\omega)|\langle L_2\|\mathbf{r}\|L\rangle|^2(-1)^{L+L_2}\begin{Bmatrix} 1 & 1 & 2 \\ L & L & L_2 \end{Bmatrix} G(L)_2 1/6^{1/2}\langle T(L)_{20}^{\dagger}\rangle$$

$$\times (3\cos^2\theta - 1) \tag{6.2.1a}$$

$$I\eta_3(\mathbf{n}) = C(\omega)|\langle L_2\|\mathbf{r}\|L\rangle|^2(-1)^{L+L_2}\begin{Bmatrix} 1 & 1 & 2 \\ L & L & L_2 \end{Bmatrix} G(L)_2\left(\frac{3}{2}\right)^{1/2}$$

$$\times \langle T(L)_{20}^{\dagger}\rangle\sin^2\theta \tag{6.2.1b}$$

The monopole can be obtained from a measurement of the total cross section. In order to determine $\langle T(L)_{20}^{\dagger}\rangle$ experimentally either the angular distribution I or the parameter $I\eta_3$ can be measured. Usually, a combination of both parameters is determined. The emitted radiation is detected at right angles to the incident beam axis. We will take the incoming beam direction as the Z axis of our coordinate system and the direction of observation $\mathbf{n}$ as the X axis. The unit vectors $\mathbf{e}_1$ and $\mathbf{e}_2$, specifying the detector system introduced in Section 6.1, are then parallel to $-Z$ and Y, respectively.

The radiation detected in the X direction passes a Nicol prism which has its axis of transmission at an angle β to the incident beam direction (this corresponds to an angle $180° - \beta$ with respect to the axis $\mathbf{e}_1$). The transmitted light is linearly polarized with polarization vecor $\mathbf{e}$ given by Eq. (1.2.5) with $180° - \beta$ replacing β (see Figure 6.2, which should be compared

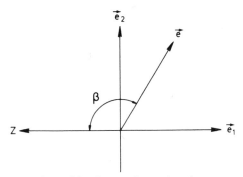

Figure 6.2. See text for explanations.

with Figure 1.4):

$$\mathbf{e} = -\mathbf{e}_1 \cos \beta - \mathbf{e}_2 \sin \beta \qquad (6.2.2)$$

The transmitted intensity is obtained from Eq. (1.2.29) with $\eta_1 = \eta_2 = 0$:

$$I_e = (1/2)[I(X) + I\eta_3(X) \cos 2\beta] \qquad (6.2.3)$$

where $I(X)$ and $I\eta_3(X)$ are given by Eqs. (6.2.1a) and (6.2.1b), respectively, with $\theta = 90°$, $\varphi = 0°$.

Usually one determines experimentally the *polarization P*, which is defined as

$$\blacktriangleright \qquad P = \frac{I_\parallel - I_\perp}{I_\parallel + I_\perp} \qquad (6.2.4)$$

Here, $I_\parallel$ and $I_\perp$ denote the intensities of the emitted light which has passed through the Nicol prism with its axis of transmission respectively parallel $(\beta = 0°)$ and perpendicular $(\beta = 90°)$ to the Z axis. Using Eq. (6.2.3) we obtain

$$P = \frac{I\eta_3(X)}{I(X)} = \eta(X)_3 \qquad (6.2.5)$$

Since $\eta_1 = \eta_2 = 0$ the magnitude of the parameter (6.2.4) is equal to the degree of coherence $P = (\eta_1^2 + \eta_2^2 + \eta_3^2)^{1/2}$ introduced in Section 1.2. Specializing Eqs. (6.2.1) to $\mathbf{n}$ parallel to X, that is, by inserting $\theta = 90°$, $\varphi = 0$ we obtain

$$P = \frac{(-1)^{L+L_2}\begin{Bmatrix} 1 & 1 & 2 \\ L & L & L_2 \end{Bmatrix} G(L)_2 (3/2)^{1/2} \langle T(L)_{20}^\dagger \rangle}{\dfrac{1}{\gamma}\dfrac{2}{3(2L+1)^{1/2}}\langle T(L)_{00}\rangle + (-1)^{L+L_2}\begin{Bmatrix} 1 & 1 & 2 \\ L & L & L_2 \end{Bmatrix} G(L)_2 \dfrac{1}{6^{1/2}}\langle T(L)_{20}^\dagger \rangle} \qquad (6.2.6)$$

Thus combining a measurement of P with a determination of the total cross section Q enables the alignment parameter to be extracted from the experimental results.

For singlet–singlet transitions $G(L)_K = 1/\gamma$ and Eq. (6.2.6) reduces to the expression

$$P = \frac{(-1)^{L+L_2}\begin{Bmatrix} 1 & 1 & 2 \\ L & L & L_2 \end{Bmatrix}\left(\dfrac{3}{2}\right)^{1/2} \langle T(L)_{20}^\dagger \rangle}{\dfrac{2}{3(2L+1)^{1/2}}\langle T(L)_{00}\rangle + (-1)^{L+L_2}\begin{Bmatrix} 1 & 1 & 2 \\ L & L & L_2 \end{Bmatrix}\dfrac{1}{6^{1/2}}\langle T(L)_{20}^\dagger \rangle} \qquad (6.2.7)$$

As an illustration consider the case of radiation emitted in a $^1D \rightarrow {}^1P$

transition. The alignment parameter may be expressed in terms of the total cross sections $Q(M)$ using Eq. (4.6.11):

$$\langle T(2)_{20}^\dagger \rangle = 5^{1/2} \sum_M (-1)^M \begin{pmatrix} 2 & 2 & 2 \\ M & -M & 0 \end{pmatrix} Q(M)$$

$$= -(2/7)^{1/2}[2Q(2) - Q(1) - Q(0)] \qquad (6.2.8a)$$

where the relation (4.6.9) has been used. Similarly, from Eq. (4.6.10)

$$\langle T(2)_{00} \rangle = (1/5)^{1/2}[2Q(2) + 2Q(1) + Q(0)] \qquad (6.2.8b)$$

Using Eqs. (6.2.8) together with the numerical value of the $6j$ symbol in Eq. (6.2.7) yields

$$P = \frac{3[-2Q(2) + Q(1) + Q(0)]}{6Q(2) + 9Q(1) + 5Q(0)} \qquad (6.2.9)$$

6.6.2. Threshold and Pseudothreshold Excitation

As was first discussed by Percival and Seaton (1957) the formulas given above simplify considerably for threshold excitation. Since all spin couplings have been neglected during the collision orbital and spin angular momenta are separately conserved, which gives in particular

$$M_0 + m_0 = M + m_1$$

where M_0 and m_0 (M and m_1) are the magnetic quantum numbers of the initial (final) atoms and electrons, respectively. The incident electron has no component of its orbital angular momentum along its direction of propagation ($m_0 = 0$). After the excitation at threshold the projectile electron has zero energy and hence zero orbital angular momentum ($m_1 = 0$). It follows that the magnetic quantum number of the atoms cannot change during the collision and, since excitation from the groundstate with $L_0 = 0$ is assumed, only the substate with $M = 0$ can be excited at threshold.

As a consequence only the cross section $Q(0)$ is nonzero at threshold and from Eq. (4.6.11):

$$\langle T(L)_{20}^\dagger \rangle = 5^{1/2}(-1)^L \begin{pmatrix} L & L & 2 \\ 0 & 0 & 0 \end{pmatrix} Q(0) \qquad (6.2.10a)$$

Similarly,

$$\langle T(L)_{00} \rangle = Q(0)/(2L + 1)^{1/2} \qquad (6.2.10b)$$

By inserting Eqs. (6.2.10) into Eq. (6.2.6) it can be seen that $Q(0)$ cancels in the resulting expression. *The threshold polarization P_{thr} is therefore* *a quantity which depends only on the geometry of the excitation process independent of any (calculated or measured) cross section value:*

$$P_{thr} = \frac{\left(\dfrac{15}{2}\right)^{1/2}(-1)^{L_2}G(L)_2\begin{Bmatrix}1&1&2\\L&L&L_2\end{Bmatrix}\begin{pmatrix}L&L&2\\0&0&0\end{pmatrix}}{\dfrac{1}{\gamma}\dfrac{2}{3(2L+1)}+\left(\dfrac{5}{6}\right)^{1/2}(-1)^{L_2}G(L)_2\begin{Bmatrix}1&1&2\\L&L&L_2\end{Bmatrix}\begin{pmatrix}L&L&2\\0&0&0\end{pmatrix}}$$

(6.2.11)

Specializing Eq. (6.2.11) to the case of $L = 1$, $L_2 = 0$, $S_1 = 0$ gives $P_{thr} = 1$. This is readily understood by noting that the excited atoms are in a pure state $|L, M = 0\rangle$ immediately after the excitation. The detected photons are thus emitted in a transition between two pure states $|L, M = 0\rangle \rightarrow |0\rangle$ and hence are necessarily completely polarized. If $L_2 \neq 0$ the emitted radiation is depolarized since the final atomic states (with $M_2 = \pm 1, 0$) are not detected. The corresponding depolarization effect is described explicitly by the $6j$ symbol in Eq. (6.2.11). If, in addition, $S_1 \neq 0$ a further depolarization is observed, caused by the coupling to the undetected spin system, which is described by the factor $G(L)_2$ in Eq. (6.2.11). In general $G(L)_2$ is given by Eq. (5.5.3). If the fine-structure levels do not overlap Eq. (5.5.8) applies and the initial coherence between different fine-structure states is destroyed. If hyperfine-structure interaction must also be taken into account then the relevant perturbation factor can be obtained from Eq. (4.7.22). This discussion shows once more the importance of fine (and hyperfine) interaction and the effect of a finite level width, both of which can considerably affect polarization. An interpretation of these results in terms of the vector model is given in the review by Kleinpoppen (1969).

There is considerable interest in the polarization of impact radiation at threshold. Attempts at direct measurement of threshold polarization have been limited because of intensity problems and also because of the effects of cascades and resonances in the energy range just above threshold. It has been pointed out by King *et al.* (1972) that polarization measurements made for the subensemble of atoms which have been excited by forward-scattered electrons reproduces threshold conditions as far as polarization is concerned and that errors due to cascade and resonance effects are eliminated. In fact, forward-scattered electrons have a zero angular momentum component both before and after the scattering ($m_0 = m_1 = 0$) with respect to the direction of motion as the quantization axis. In this case exciting the atoms from the ground state (with $L = 0$) and neglecting spin–orbit effects during the collision enables only magnetic substates with $M = 0$ to be excited. The relevant multipoles $\langle T_{00} \rangle$ and $\langle T_{20}^{\dagger} \rangle$ are given by Eqs. (6.2.10) with $Q(0)$

replaced by the differential cross section $\sigma(0)$ describing excitation of the $M = 0$ substates by forward-scattered electrons. Forward-scattered electrons and emitted photons are then detected in coincidence and P measured. $\sigma(0)$ cancels in the expression for P and P is therefore given by Eq. (6.2.11).

This technique has been recently applied in atomic and molecular physics [see, for example, the report by McConkey (1980), and references therein].

6.3. Effect of a Weak Magnetic Field

6.3.1. Perturbation Coefficients for Various Geometries. Coherence Phenomena

In this section we will consider the effect of a magnetic field on light emission. The field is assumed to be weak, that is, the mean value of the magnetic interaction is assumed to be much smaller than the separation of the relevant zero field levels. With this assumption the theory developed in Section 4.7.4 can be used and the effect of the field on the excitation process can be neglected but must be allowed for in the description of the time evolution of the excited states between excitation and decay. In terms of the vector model disturbance due to the field is described by the precessional motion of the angular momentum vectors around the direction of the field $\mathbf{H}$ at the Larmor frequency ω_L of the excited states.

Assuming that states $|JM\rangle$ have been excited at $t = 0$ the polarization density matrix of the emitted photons is obtained by specializing Eq. (5.4.2) to the case under discussion:

$$\dot{\rho}(\mathbf{n}, t)_{\lambda'\lambda} = C(\omega) \sum_{KQqq'} \text{tr}\,[r_{-\lambda'} T(J_1)_{Kq} r^{\dagger}_{-\lambda}] D(0\theta\varphi)^{(K)}_{qq'} G(t)^{Qq'}_{KK}$$

$$\times \exp(-\gamma t) \langle T(J_1)^{\dagger}_{KQ} \rangle \tag{6.3.1}$$

with $q = \lambda' - \lambda$ and where the trace is given by Eq. (5.2.4). The relevant perturbation coefficient is given by Eq. (4.7.30):

$$G(t)^{Qq'}_{KK} = \sum_{Q'} D(0\beta'\alpha')^{(K)}_{Q'q'} \exp(-i\omega_L Q't) D(0\beta'\alpha')^{(K)}_{Q'Q} \tag{6.3.2}$$

where β' and α' denote the polar angles of the field direction $\mathbf{H}$ in the coordinate system XYZ defined by the excitation process. We will now derive explicit expressions for the perturbation coefficients for some geometries of interest.

6.3.1.1. Field Parallel to Z

In this case Eq. (4.7.31) applies:

$$G(t)_{KK}^{Qq'} = \exp(-i\omega_L tQ)\delta_{Qq'} \tag{6.3.3}$$

and Eq. (6.3.1) reduces to the expression

$$\dot{\rho}(\mathbf{n}, t)_{\lambda'\lambda} = C(\omega) \sum_{KQ} \text{tr}\,[r_{-\lambda'}T(J_1)_{K_1\lambda'-\lambda}r_{-\lambda}^{\dagger}]D(0\theta\varphi)_{\lambda'-\lambda,Q}^{(K)}$$

$$\times \exp(-i\omega_L Qt - \gamma t)\langle T(J_1)_{KQ}^{\dagger}\rangle \tag{6.3.4}$$

This equation shows that the angular distribution and polarization of the emitted radiation oscillate as a function of the magnetic field strength. The quantum beats appear when $Q \neq 0$, that is, when *different sublevels have been coherently excited*. The magnetic field will have no effect if the excitation process is axially symmetric with respect to Z (see, for example, the case discussed in Section 4.5.3).

*6.3.1.2. **n** Parallel to X, **H** Parallel to Y*

Next we consider the situation in which the emitted light is observed in the X direction of the collision system and the field is directed along the Y axis. In this case $\theta = 90°$, $\varphi = 0$, and $\beta' = \alpha' = 90°$ in Eqs. (6.3.1) and (6.3.2).

After some algebraic manipulations we obtain

$$\sum_{q'} D(0, \pi/2, 0)_{\lambda'-\lambda,q'}^{(K)} G(t)_{KK}^{Qq'} = \exp(i\pi Q/2)(-1)^K d(\omega_L t)_{-\lambda'+\lambda,Q}^{(K)} \tag{6.3.5}$$

Substitution of Eq. (6.3.5) into Eq. (6.3.1) yields

$$\dot{\rho}(X,t)_{\lambda'\lambda} = C(\omega) \sum_{KQ} \text{tr}[r_{-\lambda'}T(J_1)_{K,\lambda'-\lambda}r_{-\lambda}^{\dagger}]\exp(i\pi Q/2)(-1)^K d(\omega_L t)_{-\lambda'+\lambda,Q}^{(K)}$$

$$\times \exp(-\gamma t)\langle T(J_1)_{KQ}^{\dagger}\rangle \tag{6.3.6}$$

The time modulation of $\dot{\rho}_{\lambda'\lambda}$ is given by the factor $d(\omega_L t)_{-\lambda'+\lambda,Q}^{(K)}$.

As an example of the application of Eq. (6.3.5) consider an atomic ensemble excited in a process axially symmetric with respect to the Z axis (for example, excitation by a beam of unpolarized light or in beam–foil excitation with the foil axis parallel to the incident beam axis). The excited ensemble is then characterized in terms of a monopole and an alignment parameter with $K = 2$, $Q = 0$. The intensity $I(x, t)$ observed at time t in the

X direction is given by

$$I(X, t) = \dot{\rho}(X, t)_{11} + \dot{\rho}(X_1 t)_{-1-1}$$

$$= 2C(\omega) \operatorname{tr} (r_{-1} r^{+}_{-1}) \exp (-\gamma t) \langle T(J_1)_{00} \rangle / (2J_1 + 1)^{1/2}$$

$$+ C(\omega) \operatorname{tr} [r_{-1} T(J_1)_{20} r^{+}_{-1}](1/2)(1 + 3 \cos 2\omega_L t) \exp (-t) \langle T(J_1)^{+}_{20} \rangle$$

$$(6.3.7)$$

where explicit expressions for the d functions together with

$$\operatorname{tr} [r_{-\lambda'} T(J_1)_{Kq} r^{+}_{-\lambda}] = (-1)^K \operatorname{tr} [r_{\lambda'} T(J_1)_{K-q} r^{+}_{\lambda}]$$

have been used. Equation (6.3.7) shows that the intensity $I(X, t)$ exhibits oscillations with twice the Larmor frequency. It is instructive to consider the coherence effect responsible for these quantum beats. Only $\langle T_{00} \rangle$ and $\langle T^{+}_{20} \rangle$ contribute to Eq. (6.3.7) as a result of the *incoherent* excitation of the substates $|J_1 M_1 \rangle$ with different M_1, where M_1 is defined with respect to Z as the quantization axis. The interference effects between the eigenstates $|J_1 M'_1 \rangle$ of the Hamiltonian

$$H = H_0 - g\mu_B \mathbf{JH}$$

which governs the time evolution between excitation and decay are responsible for the quantum beats (see the discussion in Section 5.4.2). Defining M'_1 with respect to $\mathbf{H}$ as the quantization axis we have

$$\exp \left(-\frac{i}{\hbar} Ht \right) |J'_1 M'_1 \rangle = \exp \left[-\frac{i}{\hbar} (E_1 - \hbar \omega_L M'_1) t \right] |J_1 M'_1 \rangle$$

Any state $|J_1 M_1 \rangle$ can be written as a linear superposition of states $|J_1 M'_1 \rangle$:

$$|J_1 M_1 \rangle = \sum_{M'_1} a (M'_1 M_1) |J_1 M'_1 \rangle$$

(where not all possible values of M'_1 may exist in the new frame). The density matrix ρ describing the excited atoms is diagonal in M but in general nondiagonal in M'_1 provided $\langle T_{20} \rangle$ is different from zero (otherwise ρ is proportional to the identity matrix which is diagonal in any representation). This coherence between the substates gives rise to interference effects expressed by the d function in Eq. (6.3.7). This example illustrates once again that an excitation process which is incoherent for one quantization axis may be coherent for a different axis.

In general, if the magnetic field is not parallel to Z, interference terms will occur, even if different substates $|J_1 M_1 \rangle$ are incoherently excited. In order to observe quantum beats it is sufficient to produce a different population of the states $|J_1 M_1\rangle$, that is, a nonvanishing alignment parameter as shown by Eq. (6.3.7).

Equation (6.3.7) can be applied to a determination of the alignment parameter as well as the gyromagnetic ratio [further details on this can be found in the review by Macek and Burns (1976)].

6.3.1.3. n Parallel to X, H Parallel to X

Finally we will consider a geometry in which the direction of observation n and the field are parallel and directed along the X axis. In this case it can be shown that

$$\sum_{q'} D\left(0, \frac{\pi}{2}, 0\right)^{(K)}_{\lambda'-\lambda, q'} G(t)^{Qq'}_{KK} = d\left(-\frac{\pi}{2}\right)^{(K)}_{Q_1 \lambda'-\lambda} \exp\left[-i\omega_L(\lambda'-\lambda)t\right] \quad (6.3.8)$$

Substitution of this expression into Eq. (6.3.1) yields

$$\dot{\rho}(x, t)_{\lambda'\lambda} = C(\omega) \sum_{KQ} \text{tr} \left[r_{-\lambda'} T(J_1)_{K,\lambda'-\lambda} r^{\dagger}_{-\lambda}\right] d\left(-\frac{\pi}{2}\right)^{(K)}_{Q,\lambda'-\lambda}$$
$$\times \exp\left[-i\omega_L(\lambda'-\lambda)t - \gamma t\right]\langle T(J_1)^{\dagger}_{KQ}\rangle \quad (6.3.9)$$

It should be noted that in this geometry the interference effects are independent of Q (and therefore independent of whether substates with different M_1 have been coherently excited or not). The quantum beats depend only on $\lambda' - \lambda$ and only the off-diagonal elements of the polarization density matrix will exhibit time modulations. We will consider Eq. (6.3.9) and its consequences in detail in the following section.

6.3.2. Magnetic Depolarization. Theory of the Hanle Effect

In the 1920s the depolarization of resonance fluorescence from atoms subjected to external magnetic fields was discovered by Hanle (1924). During the last 30 years the techniques of magnetic depolarization have been further developed and widely applied to study Zeeman and hyperfine structure of excited and ground states and to determine radiative lifetimes and interatomic relaxation rates. In this section we will derive the basic formulas necessary for the description of such experiments.

Consider an atomic system excited by linearly polarized light. It is convenient to define the "collision" system in the following way. The Z axis is chosen to be parallel to the polarization vector e of the incident light, the Y axis parallel to the incident beam axis. We will consider a geometry in which the emitted resonance light is observed in the X direction with the magnetic field parallel to X.

As discussed in Section 4.5.3 absorption of plane-polarized light with **e** parallel to Z will produce alignment but no orientation in the excited atoms. The atomic system can therefore be completely characterized by the two parameters $\langle T(J_1)_{00} \rangle$ and $\langle T(J_1)_{20}^\dagger \rangle$. The polarization density matrix of the emitted radiation is given by Eq. (6.3.9) for the geometry under discussion. The Stokes parameters can be calculated from Eqs. (6.3.9) and (1.2.23) using the detector system defined in Section 6.2.1 (with **n** parallel to X, $\mathbf{e}_1$ parallel to $-Z$, and $\mathbf{e}_2$ parallel to Y). Assuming that the excitation time ($t = 0$) is sharply defined we obtain

$$I(x, t) = \frac{2C(\omega)}{(2J_1 + 1)^{1/2}} \, \text{tr} \, (r_{-1} r_{-1}^\dagger) \exp (-\gamma t) \langle T(J_1)_{00} \rangle$$

$$- C(\omega) \, \text{tr} \, [r_{-1} T(J_1)_{20} r_{-1}^\dagger] \exp (-\gamma t) \langle T(J_1)_{20}^\dagger \rangle \quad (6.3.10a)$$

$$I\eta_3(x, t) = -C(\omega) \, \text{tr} \, [r_{-1} T(J_1)_{2-2} r_{+1}^\dagger] \left(\frac{3}{2}\right)^{1/2} (\cos 2\omega_L t)$$

$$\times \exp (-\gamma t) \langle T(J_1)_{20}^\dagger \rangle \quad (6.3.10b)$$

$$I\eta_1(x, t) = C(\omega) \, \text{tr} \, [r_{-1} T(J_1)_{2-2} r_{+1}^\dagger] \left(\frac{3}{2}\right)^{1/2} (\sin 2\omega_L t)$$

$$\times \exp (-\gamma t) \langle T(J_1)_{20}^\dagger \rangle \quad (6.3.10c)$$

$$I\eta_2(x, t) = 0 \quad (6.3.10d)$$

The last equation is a consequence of the fact that the initially excited atoms were unoriented and that the magnetic field does not mix multipoles with different rank [see Eq. (6.3.9)]. Note that $I(X, t)$ does not depend on the field.

The time-integrated Stokes parameters are of particular interest. Integrating Eqs. (6.3.10) over a time interval $0 \cdots t_R$ where t_R is much larger than the mean lifetime (so that the upper integration limit can be extended to infinity with negligible error) we obtain

$$I(X) = \frac{2C(\omega)}{\gamma(2J_1 + 1)^{1/2}} \, \text{tr} \, (r_{-1} r_1^\dagger) \langle T(J_1)_{00} \rangle$$

$$- \frac{C(\omega)}{\gamma} \, \text{tr} \, [r_{-1} T(J_1)_{20} r_{-1}^\dagger] \langle T(J_1)_{20}^\dagger \rangle \quad (6.3.11a)$$

$$I\eta_3(X) = -\frac{\gamma}{\gamma^2 + 4\omega_L^2} \left(\frac{3}{2}\right)^{1/2} C(\omega) \, \text{tr} \, [r_{-1} T(J_1)_{2-2} r_{+1}^\dagger] \langle T(J_1)_{20}^\dagger \rangle \quad (6.3.11b)$$

$$I\eta_1(X) = \frac{2\omega_L}{\gamma^2 + 4\omega_L^2} \left(\frac{3}{2}\right)^{1/2} C(\omega) \, \text{tr} \, [r_{-1} T(J_1)_{2-2} r_{+1}^\dagger] \langle T(J_1)_{20}^\dagger \rangle \quad (6.3.11c)$$

If the fluorescence light observed in the X direction passes through a linear polarizer with the axis of transmission at an angle β to the Z axis then the polarization vector $\mathbf{e}$ of the transmitted light is given by Eq. (6.2.2), which is a special case of the general equation (1.2.4) with $\delta = 0$. The transmitted intensity can be obtained by substituting Eqs. (6.3.11) into Eq. (1.2.29) with $\delta = 0$:

$$I(X)_e = I(X)/2 - (1/2)(3/2)^{1/2} \, \text{tr} \, [\mathbf{r}_{-1} \, T(J_1)_{2-2} r_1^\dagger] \langle T(J_1)_{20} \rangle$$

$$\times \left(\frac{\gamma}{\gamma^2 - 4\omega_L^2} \cos 2\beta + \frac{2\omega_L}{\gamma^2 + 4\omega_L^2} \sin 2\beta \right) \qquad (6.3.12)$$

It is instructive to discuss an example of the use of Eq. (6.3.12). Consider the case in which the incident linearly polarized light excites atoms in a ^1S ground state to a ^1P state. Because of the dipole selection rules only the substate with magnetic quantum number $M = 0$ can be excited and $\langle M = 0 | \rho | M = 0 \rangle \equiv \rho_{00}$ is the only nonvanishing element of the excited state density matrix. Application of Eq. (4.3.3) gives

$$\langle T_{00} \rangle = \frac{1}{3^{1/2}} \rho_{00}, \qquad \langle T_{20} \rangle = -\left(\frac{2}{3} \right)^{1/2} \rho_{00} \qquad (6.3.13)$$

Calculating the traces in Eqs. (6.3.11a) and (6.3.12) and describing the subsequent decay to the atomic ground state by means of Eq. (5.2.4) we obtain

$$\text{tr} \, (r_{-1} r_{-1}^\dagger) = (1/3) |\langle 0 \| \mathbf{r} \| 1 \rangle|^2$$

$$\text{tr} \, [r_{-1} T(1)_{20} r_{-1}^\dagger] = [1/3(6)^{1/2}] |\langle 0 \| \mathbf{r} \| 1 \rangle|^2$$

$$\text{tr} \, [r_{-1} T(1)_{2-2} r_1^\dagger) = (1/3) |\langle 0 \| \mathbf{r} \| 1 \rangle|^2 \qquad (6.3.14)$$

Substitution of Eqs. (6.3.13) and (6.3.14) into the expressions (6.3.11a) and (6.3.12) gives

$$I(X) = C(\omega) |\langle 0 \| \mathbf{r} \| 1 \rangle|^2 \rho_{00}/3 \qquad (6.3.15a)$$

and

$$I(X)_e = [I(X)/2] \left[1 + \frac{\gamma^2}{\gamma^2 + 4\omega_L^2} \cos 2\beta + \frac{2\omega_L \gamma}{\gamma^2 + 4\omega_L^2} \sin 2\beta \right] \qquad (6.3.15b)$$

It should be noted that *the shape of the observed signal* $I(X)_e$ described by Eq. (6.3.15) *depends on the orientation of the polarizer in the detection beam*. The shape is called a *Lorentzian shape* for $\beta = 0$ and a *dispersion shape* for $\beta = 45°$. Figures (6.3a) and (6.3b) show the curves obtained in the special case when Eq. (6.3.15) applies.

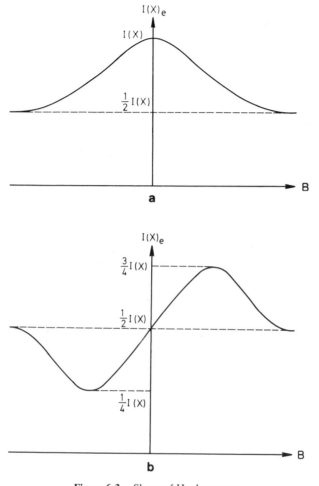

Figure 6.3. Shape of Hanle curves.

Finally we will consider the polarization P defined by Eq. (6.2.4) as

$$P = \frac{I_{\parallel} - I_{\perp}}{I_{\parallel} + I_{\perp}} \qquad (6.3.16)$$

where $I_{\parallel}$ ($I_{\perp}$) is the intensity transmitted by the polarizer if its axis of transmission is parallel (perpendicular) to Z. $I_{\parallel}$ and $I_{\perp}$ can be obtained by inserting $\beta = 0°$ and $\beta = 90°$ in Eq. (1.2.29), which gives

$$P = \eta_3 \qquad (6.3.17)$$

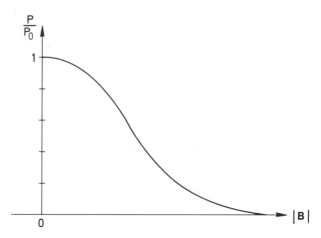

Figure 6.4. Depolarization of resonance fluorescence.

Substitution of Eqs. (6.3.11a) and (6.3.11b) into Eq. (6.3.17) yields

$$P = P_0 \frac{\gamma^2}{\gamma^2 + 4\omega_L^2}$$

where P_0 denotes the polarization detected in the field-free case ($\omega_L = 0$). The factor

$$\frac{\gamma^2}{\gamma^2 + 4\omega_L^2} < 1$$

describes the *depolarization* of the emitted fluorescence light caused by the magnetic field. If **H** is slowly varied the polarization P of the radiation changes from its maximum possible value at the zero field to steadily decreasing values with increasing field strength (Figure 6.4). This constitutes the *Hanle effect* or the *magnetic* depolarization of resonance radiation.

The emphasis of our discussions here has been to illustrate how Eqs. (6.3.12) and (6.3.15) are direct consequences of the general theory presented in Chapters 4 and 5. For an interpretation of the Hanle effect in terms of a semiclassical model, experimental results and applications we refer, for example, to the review by Cohen-Tannoudji and Kastler (1966) and the book by Corney (1977).

7

Quantum Theory of Relaxation

7.1. Density Matrix Equations for Dissipative Quantum Systems

7.1.1. Conditions of Irreversibility. Markoff Processes

Consider a system which is not closed but in continuous contact with its surroundings, exchanging energy, polarization and so forth. If initially the system is in a nonequilibrium state then—under certain conditions which will be specified below—it will at some later time go over into an equilibrium state determined by external conditions such as temperature. This gradual evolution into an equilibrium state is called a *relaxation process*. In the present chapter we will consider some methods for studying processes of this kind.

Relaxation phenomena are irreversible processes. The fundamental quantum mechanical equations of motion, the Schrödinger and Liouville equations, describe a reversible evolution in the course of time, and hence a major problem is that of the solution of the question how irreversibility can arise if the behavior of microscopic particles is strictly reversible. In recent years there has been success in answering this question. A detailed treatment of the modern theory is outside the scope of this book and the reader is referred to modern textbooks on statistical physics for a more detailed account, for example to the book by Prigogine (1981).

We will start with the concepts introduced in Section 3.2. Consider a system S interacting with an unobserved system R. We will denote the density matrix characterizing the total system by $\rho(t)$ and the total Hamiltonian by $H = H_S + H_R + V$, where H_S and H_R are the Hamiltonians for the uncoupled systems S and R, respectively, and V describes the interactions between S and R. In the interaction picture the time evolution of $\rho(t)$ is given by Eq. (2.4.41) or, alternatively, by Eq. (2.4.42). Inserting Eq.

(2.4.42) back into Eq. (2.4.41) gives

$$\dot{\rho}(t)_I = -(i/\hbar)[V(t)_I, \rho(0)] - (1/\hbar)^2 \int_0^t dt' \, [V(t)_I, [V(t')_I, \rho(t')_I]] \quad (7.1.1)$$

where $\dot{\rho}(t)_I$ is the time derivation of $\rho(t)_I$, and $\rho(t)_I$ and $V(t)_I$ are operators in the interaction picture which are related to their Schrödinger picture counterparts by Eqs. (2.4.37) and (2.4.25), respectively, with H_0 replaced by $H_S + H_R$.

The reduced density matrix $\rho(t)_S$, describing the system of interest S, is obtained from $\rho(t)$ by taking the trace over all variables of the unobserved system R according to Eq. (3.2.5). Hence, in the interaction picture,

$$\rho(t)_{SI} = \text{tr}_R \, \rho(t)_I \quad (7.1.2)$$

and from Eq. (7.1.1)

$$\dot{\rho}(t)_{SI} = -(i/\hbar) \, \text{tr}_R \, [V(t)_I, \rho(0)_I]$$

$$- (1/\hbar)^2 \int_0^t dt' \, \text{tr}_R \, [V(t)_I, [V(t')_I, \rho(t')_I]] \quad (7.1.3)$$

In writing Eqs. (7.1.1) and (7.1.3) it has been assumed that the interaction is switched on at time $t = 0$. Prior to this S and R are uncorrelated and the total density matrix is given by the direct product

$$\rho(0) = \rho(0)_S \rho(0)_R = \rho(0)_I \quad (7.1.4)$$

(see Appendix A).

The coupling between the two systems may result in a reversible exchange of energy, polarization etc. An example has been discussed in Section 5.7, the coupling of orbital angular momentum to an undetected spin system. In order for an irreversible process to occur further conditions must be imposed on the unobserved system in order to prevent the energy initially in the system S from returning back from the unobserved system R to S in any finite time.

At this point we follow Fano (1957) and make the first of two key assumptions. It is assumed *that R has so many degrees of freedom that the effects of the interaction with S dissipate away quickly and will not react back onto S to any significant extent so that R remains described by a thermal equilibrium distribution at constant temperature, irrespective of the amount of energy and polarization diffusing into it from the system S.* In other words, it is assumed that the reaction of S on R is neglected [so that the R system is represented by $\rho(0)_R$ at all times] and the correlations between S and R, induced by the interaction, are neglected. In this case $\rho(t)_I$ can be replaced by the simpler density matrix

▶ $$\rho(t)_I \rightarrow \overline{\rho(t)_I} = \rho(t)_{SI}\rho(0)_R \qquad (7.1.5)$$

at any time t without introducing any significant error in the calculation of $\rho(t)_{SI}$. $\rho(0)_R$ is represented by the density matrix (2.6.1):

$$\rho(0)_R = \exp(-\beta H_R)/Z \qquad (7.1.6)$$

Equations (7.1.5) represents our *basic condition of irreversibility*.

In the following we will consider the behavior of a "small" system S, the dynamical system, coupled to a "large" system R with many degrees of freedom. Throughout this chapter we will refer to the large system as the "heat bath" or "reservoir." For example, atoms in a gas will collide with other atoms and these can act as a heat reservoir for the atoms considered. Light in a cavity is in interaction with the walls which then play the role of a heat bath for the light. In magnetic resonance experiments the spin variables interact with other degrees of freedom (the "lattice") and these other variables form the heat reservoir.

Replacing $\rho(t)_I$ in Eq. (7.1.3) by the approximate density matrix (7.1.5) gives

$$\dot{\rho}(t)_{SI} = -(i/\hbar)\,\mathrm{tr}_R\left[V(t)_I, \rho(0)_S\rho(0)_R\right]$$

$$- (1/\hbar)^2 \int_0^t dt'\,\mathrm{tr}_R\left[V(t)_I, [V(t')_I, \rho(t')_{SI}\rho(0)_R]\right] \qquad (7.1.7)$$

It should be noted that the corections neglected in Eqs. (7.1.5) and (7.1.7) can be treated systematically by successive approximations. If the interaction term V is zero then system and reservoir are uncorrelated and $\rho(t)_I = \overline{\rho(t)_I}$. If V is small (that is, $|V| \ll |H_S|, |V| \ll |H_R|$) then we can write

$$\rho(t) = \rho(t)_{SI}\rho(0)_R + \Delta\rho \qquad (7.1.8)$$

such that $\Delta\rho$ is small of order V. If Eq. (7.1.8) is inserted in the integral in Eq. (7.1.3) and only terms of order V^2 are retained then Eq. (7.1.7) is obtained. Equation (7.1.7) is therefore the equation of motion for the dynamic system up to second order in the interaction.

Equation (7.1.7) contains $\rho(t')_{SI}$ in the integral, and hence the behavior of the system depends on its past history from $t' = 0$ to $t' = t$. The motion of the system S is, however, damped by the coupling to the reservoir and damping destroys the knowledge of the past behavior of the system. We therefore make our second key assumption: $\dot{\rho}(t)_{SI}$ *depends only on* $\rho(t)_{SI}$, *its present value*. In other words, it is assumed that the system loses all memory of its past. Hence, in Eq. (7.1.7) we can make the substitution

▶ $$\rho(t')_{SI} \rightarrow \rho(t)_{SI} \qquad (7.1.9)$$

This substitution is the *Markoff approximation* and gives

▶ $$\dot{\rho}(t)_{SI} = -(i/\hbar)\, \text{tr}_R\, [V(t)_I, \rho(0)_S\rho(0)_R]$$

$$-(1/\hbar)^2 \int_0^t dt'\, \text{tr}_R\, [V(t)_I, [V(t')_I, \rho(t)_{SI}\rho(0)_R]] \qquad (7.1.10)$$

We will consider the Markoff approximation in more detail in the following section.

7.1.2. Time Correlation Functions. Discussion of the Markoff Approximation

The next step in the analysis of Eq. (7.1.9) is to consider the coefficients of $\rho(t)_{SI}$ in this equation. The development here follows that of Loisell (1973) to whom reference should be made for more detailed treatment of specific points (see also Sargent, Scully, and Lamb, 1974; Haken, 1970).

Let us assume that the interaction operator can be written in the form

$$V = \sum_i Q_i F_i \qquad (7.1.11)$$

where the F_i are reservoir operators and the Q_i are operators acting only on the variables of the dynamic system. In the interaction picture

$$V(t)_I = \exp\,[i(H_S + H_R)t/\hbar]V \exp\,[-i(H_S + H_R)t/\hbar]$$

$$= \sum_i F(t)_i Q(t)_i \qquad (7.1.13)$$

where

$$F(t)_i = \exp\,(iH_R t/\hbar)F_i \exp\,(-iH_T t/\hbar) \qquad (7.1.13a)$$

and

$$Q(t)_i = \exp\,(iH_S t/\hbar)Q_i \exp\,(-iH_S t/\hbar) \qquad (7.1.13b)$$

Inserting the expression (7.1.13) into Eq. (7.1.10) using the fact that the operators F_i and Q_i commute and using the cyclic property of the trace gives

$$\dot{\rho}(t)_{SI} = -(i/\hbar)\sum_i \{Q(t)_i\rho(0)_{SI}\, \text{tr}_R\, (F(t)_i\rho(0)_R)$$

$$- \rho(0)_{SI}Q(t)_i\, \text{tr}_R\, (F(t)_i\rho(0)_R)\}$$

$$-(1/\hbar)^2 \sum_{ij} \int_0^t dt'\, \{(Q(t)_i Q(t')_j\rho(t)_{SI} - Q(t')_j\rho(t)_{SI}Q(t)_i)\, \text{tr}_R\, (F(t)_i F(t')_j\rho(0)_R)$$

$$-(Q(t)_i\rho(t)_{SI}Q(t')_j - \rho(t)_{SI}Q(t')_j Q(t)_i)\, \text{tr}_R\, (F(t')_j F(t)_i\rho(0)_R)\}$$

$$(7.1.14)$$

Consider first the expectation values

$$\langle F(t)_i \rangle = \text{tr}_R \left(F(t)_i \rho(0)_R \right)$$

$$= \sum_N \langle N | F(t)_i | N \rangle \langle N | \rho(0)_R | N \rangle \tag{7.1.15}$$

where the trace has been conveniently expressed in terms of eigenstates $|N\rangle$ of H_R so that the equilibrium density matrix (7.1.6) is diagonal in this representation. Assuming that the interaction operators F_i have no diagonal elements in this representation (since otherwise the free Hamiltonian could be redefined to include those parts) we then have

$$\langle F(t)_i \rangle = 0 \tag{7.1.16}$$

This is equivalent to the assumption that the interaction does not produce an average frequency shift. It then follows that the first term in Eq. (7.1.14) vanishes.

Next consider the functions

$$\langle F(t)_i F(t')_j \rangle = \text{tr}_R \left(F(t)_i F(t')_j \rho(0)_R \right) \tag{7.1.17}$$

These are *time correlation functions*, that is, expectation values of products of physical quantities taken at different times, which characterize the correlation which exists on average between interactions occurring at times t and t'. Since the reservoir is assumed to be large and such that it quickly dissipates the effects of the interaction it is expected that the reservoir will quickly "forget" its interactions with the system S. Thus it is expected that $\langle F(t)_i F(t')_j \rangle$ will be nonzero for some time interval $t - t' \gtrsim \tau$, where τ is typical of the reservoir and is called the *correlation time* of the reservoir. Interactions at times t and t' become progressively less correlated for $t - t' > \tau$ and become uncorrelated for $t - t' \gg \tau$ in which case, using Eq. (7.1.16),

$$\langle F(t)_i F(t')_j \rangle \approx \langle F(t)_i \rangle \langle F(t')_j \rangle \approx 0 \tag{7.1.18}$$

The correlation function $\langle F(t)_i F(t')_j \rangle$ is therefore a maximum at $t = t'$ and decreases with increasing $t - t'$.

The correlation time τ is a measure of the time during which, on average, some memory of the interaction is retained. The nature of τ depends on the nature of the reservoir. In the case of gases, for example, τ may be given by the mean time between two collisions. Similarly, in magnetic resonance experiments any nuclei will interact with the magnetic moment of the neighboring nuclei, and, in the case of liquids, τ is given by the mean time for which a given pair of nuclei is near to each other before diffusing away.

Finally, we note that the correlation functions (7.1.17) are *stationary*, that is, they depend only on the time difference $t - t'$. This can be shown from Eq. (7.1.13a) by using the cyclic property of the trace and the fact that the equilibrium density matrix (7.1.6) commutes with H_R:

$$\langle F(t)_i F(t')_j \rangle = \text{tr}_R \left[\exp \left(iH_R t/\hbar \right) F_i \exp \left(-iH_R t/\hbar \right) \exp \left(iH_R t'/\hbar \right) F_j \right.$$
$$\left. \times \exp \left(-iH_R t'/\hbar \right) \rho(0)_R \right]$$
$$= \text{tr}_R \left\{ \exp \left[iH_R (t - t')/\hbar \right] F_i \exp \left[-iH_R (t - t')/\hbar \right] F_j \rho(0)_R \right\}$$
$$= \langle \dot{F}(t - t')_i F_j \rangle \tag{7.1.19}$$

We will now use these results and reconsider the Markoff approximation. Because of the property (7.1.18) the integral in Eq. (7.1.7) is effectively only nonzero for a time interval $t - t' \gtrsim \tau$, that is, between times $t' \approx t - \tau$ and $t' = t$. It follows that values of $\rho(t')_{SI}$ at times t' outside this interval have little or no influence on $\dot{\rho}(t)_{SI}$ at time t. The system is therefore capable of memorizing its state for time intervals only which are not much larger than the correlation time. Usually one is interested in the macroscopic behavior of the system rather than in its detailed changes. If τ is much smaller than a characteristic time $1/\gamma$ (the damping time or decay time), required for $\rho(t)_{SI}$ to change appreciably on a macroscopic scale,

$$\tau \ll 1/\gamma \tag{7.1.20}$$

then $\rho(t')_{SI} \approx \rho(t)_{SI}$ in the integrand of Eq. (7.1.7) and the Markoff approximation holds.

Substitution of $\rho(t)_{SI}$ for $\rho(t')_{SI}$ in Eq. (7.1.7) therefore implies that we do not try to describe details of the system motion for time intervals comparable to τ. The quantity of interest is

$$\frac{\Delta \rho(t)_{SI}}{\Delta t} = \frac{\rho(t + \Delta t)_{SI} - \rho(t)_{SI}}{\Delta t} \tag{7.1.21}$$

where two values of the system density matrix are compared at times t and $t + \Delta t$, where Δt is much larger than τ but still sufficiently small that the change in $\rho(t)_{SI}$ is linear in Δt. If it is possible to find an interval Δt which satisfies this condition then $\Delta \rho / \Delta t$ can be replaced by the time derivative (7.1.10) provided that it is understood that we never use this equation to describe the changes of $\rho(t)_{SI}$ over time intervals less than τ. In this sense one often refers to the Markoff approximation as a "coarse-grained" average and the time derivative (7.1.10) is often called a "coarse-grained" derivative.

7.1.3. The Relaxation Equation. The Secular Approximation

We now return to the further development of Eq. (7.1.14). Applying the relation (7.1.19) and introducing the variable $t'' = t - t'$, $dt'' = -dt'$, the integral $\int_0^t dt' \cdots$ is transformed into the integral $\int_0^t dt'' \cdots$. The correlation function $\langle F(t'')_i F_j \rangle$ is effectively zero for $t'' \gg \tau$ and, hence, the upper integration limit can be extended to infinity with negligible error under the Markoff approximation. Using Eq. (7.1.16) we obtain

$$\dot{\rho}(t)_{SI} = -(1/\hbar)^2 \sum_{ij} \int_0^\infty dt'' \{ [Q(t)_i, Q(t - t'')_i \rho(t)_{SI}] \langle F(t'')_i F_j \rangle$$
$$- [Q(t)_i, \rho(t)_{SI} Q(t - t'')_j] \langle F_j F(t'')_i \rangle \} \tag{7.1.22}$$

It should be noted that all information on the reservoir is contained in the correlation functions. Taking matrix elements between eigenstates $|m\rangle$ of H_S and applying Eq. (7.1.13b) gives

$$\langle m | Q(t)_i | n \rangle = \exp(i\omega_{mn}t) \langle m | Q_i | n \rangle \tag{7.1.23}$$

where

$$\omega_{mn} = (E_m - E_n)/\hbar$$

Introducing the notation

$$\Gamma_{mkln}^+ = (1/\hbar)^2 \sum_{ij} \langle m | Q_i | k \rangle \langle l | Q_j | n \rangle \int_0^\infty dt'' \exp(-i\omega_{ln}t'') \langle F(t'')_i F_j \rangle \tag{7.1.24a}$$

$$\Gamma_{mkln}^- = (1/\hbar)^2 \sum_{ij} \langle m | Q_j | k \rangle \langle l | Q_i | n \rangle \int_0^\infty dt'' \exp(-i\omega_{mk}t'') \langle F_j F(t'')_i \rangle \tag{7.1.24b}$$

we obtain after some algebra

$$\langle m' | \rho(t)_{SI} | m \rangle = \sum_{n'n} \langle n' | \rho(t)_{SI} | n \rangle \left\{ -\sum_k \delta_{mn} \Gamma_{m'kkn'}^+ + \Gamma_{nmm'n'}^+ + \Gamma_{nmm'n'}^- \right.$$
$$\left. -\sum_k \delta_{n'm'} \Gamma_{nkkm}^- \right\} \exp[i(\omega_{m'n'} + \omega_{mn})t] \tag{7.1.25a}$$

which can be written in the form

$$\langle m' | \dot{\rho}(t)_{SI} | m \rangle = \sum_{n'n} \langle n' | \rho(t)_{SI} | n \rangle R_{m'mn'n} \exp[i(E_{m'} - E_m - E_{n'} + E_n)t/\hbar] \tag{7.1.25b}$$

where the t-independent parameters $R_{m'mn'n}$ are defined by the curly brackets in Eq. (7.1.25a).

In this equation the time-dependent exponential vanishes if

$$E_{m'} - E_m - E_{n'} + E_n = 0 \qquad (7.1.26)$$

Equation (7.1.25) is often approximated by the equation

$$\langle m'|\dot{\rho}(t)_{SI}|m\rangle = {\sum_{n'n}}' \langle n'|\rho(t)_{SI}|n\rangle R_{m'mn'n} \qquad (7.1.27)$$

where the prime on the summation sign indicates that only the *secular terms* are kept, that is, those terms satisfying Eq. (7.1.26). This approximation means that the "coarse-grained" derivative is taken over time intervals Δt which are long compared to a period of the free motion of the system

$$\Delta t \gg \frac{1}{\omega_{mn}}$$

so that the system goes through many cycles during Δt.

We will now consider the secular terms in more detail. Following Loisell (1973) we will consider the case in which there are no regularities in the level spacing of the system. Equation (7.1.26) is then satisfied in any of the following cases: (1) $m' = n'$, $m = n$, $m' \neq m$; (2) $m' = m$, $n' = n$, $m' \neq n'$; (3) $m' = m = n' = n$. In these cases

$$\langle m'|\dot{\rho}(t)_{SI}|m\rangle = [\langle m'|\rho(t)_{SI}|m\rangle R_{m'mm'm}]' + \delta_{m'm}{\sum_{n}}' \langle n|\rho(t)_{SI}|n\rangle R_{mmnn}$$

$$+ \delta_{m'm}\langle m'|\rho(t)_{SI}|m'\rangle R_{m'm'm'm'} \qquad (7.1.28a)$$

where the prime on the square bracket indicates that this term contributes only if $m' \neq m$ and the prime on the summation sign indicates that the term $m = n$ must be omitted. If the prime on the square bracket is dropped the third term of Eq. (7.1.28a) is automatically included, which gives

▶ $$\langle m'|\dot{\rho}(t)_{SI}|m\rangle = \delta_{m'm} \sum_{n \neq m} \langle n|\rho(t)_{SI}|n\rangle \cdot W_{mn} - \gamma_{m'm}\langle m'|\rho(t)_{SI}|m\rangle$$

$$(7.1.28b)$$

where (for $m \neq n$)

$$W_{mn} = \Gamma^+_{nmmn} + \Gamma^-_{nmmn} \qquad (7.1.29a)$$

and

$$\gamma_{m'm} = \sum_{k} (\Gamma^+_{m'kkm'} + \Gamma^-_{mkkm}) - \Gamma^+_{mmm'm'} - \Gamma^-_{mmm'm'} \qquad (7.1.29b)$$

It is left as an exercise for the reader to show that

$$(\Gamma^-_{mnkl})^* = \Gamma^+_{lknm} \tag{7.1.30}$$

from which it follows *that the parameters W_{mn} are real.*

In the approximation (7.1.28) the off-diagonal elements of the density matrix obey the equation

$$\langle m'|\rho(t)_{SI}|m\rangle = -\gamma_{m'm}\langle m'|\rho(t)_{SI}|m\rangle \tag{7.1.31}$$

The Hermiticity condition (2.2.5) implies

$$\gamma_{m'm} = \gamma^*_{mm'} \tag{7.1.32}$$

The physical importance of parameters W_{mn} and γ_{mn} will be considered in the following sections. Equation (7.1.28) is transformed into the Schrödinger picture by substituting

$$\rho(t)_{SI} = \exp{(iH_St/\hbar)}\rho(t)_S \exp{(-iH_St/\hbar)}$$

which gives

$$\blacktriangleright \quad \langle m'|\rho(t)_S|m\rangle = -(i/\hbar)\langle m'|[H_S, \rho(t)_S]|m\rangle$$
$$+ \delta_{m'm} \sum_n \langle n|\rho(t)_S|n\rangle W_{mn} - \gamma_{m'm}\langle m'|\rho(t)_S|m\rangle$$

$$\tag{7.1.33}$$

where the first term describes the motion of the unperturbed system.

The equations of motion of reduced density matrices are often called the *generalized Master equations.* Master equations were first introduced into quantum statistical physics by Pauli (1928). In their original form as used by Pauli they are rate equations for the diagonal elements of $\rho(t)_S$ (see Section 7.2). For a survey and rigorous proofs we refer to the review by Haake (1973).

Equations (7.1.25), (7.1.28), and (7.1.33) play a very important role in physical kinetics. They describe the *irreversible* behavior of a system and are therefore quite unlike the detailed equations of motion, the Schrödinger and Liouville equations. It will be useful to recall briefly the essential steps in deriving the "kinetic" equations starting with the general equation (7.1.3). The basic assumption is that the effects of the interaction between system and reservoir dissipate away quickly so that the reservoir effectively remains in thermal equilibrium and is represented by the density matrix (7.1.6). This assumption leads to Eq. (7.1.7), which is an integrodifferential equation for the elements of $\rho(t)_{SI}$. The time interval for which the integral in this relation is essentially nonzero corresponds to a correlation time τ for $V(t)_I$. If τ is small compared to a characteristic time $1/\gamma$ required for the system to change appreciably then the Markoff approximation $\rho(t')_{SI} \approx \rho(t)_{SI}$ can be applied and the upper integration limit can be extended to infinity. *The Markoff approximation enables the integrodifferential equation (7.1.7) to be reduced to a set of linear differential equations for the elements of $\rho(t)_{SI}$ with time-independent coefficients $R_{m'mn'n}$.* On retaining only the secular terms Eq. (7.1.28) is obtained.

We have presented the derivation of Eqs. (7.1.25) and (7.1.33) in some detail in order to show the assumptions made and the limits of applicability of these equations.

7.2. Rate (Master) Equations

In order to obtain an interpretation of some of the parameters occurring in Eqs. (7.1.25) and (7.1.28) we will consider the rate of change of the diagonal elements of a density matrix $\rho(t)_S$ describing a system of atoms (or nuclei) interacting with some "reservoir." Retaining only the secular terms and noting that for the diagonal elements the Schrödinger picture is equivalent to the interaction picture and using the notation

$$\langle m'|\rho(t)_S|m\rangle = \rho(t)_{m'm}$$

it is found from Eqs. (7.1.28) and (7.1.29b) that

$$\dot{\rho}(t)_{mm} = \sum_{n \neq m} \rho(t)_{nn} W_{mn} - \rho(t)_{mm} \sum_{k \neq m} (\Gamma^+_{mkkm} + \Gamma^-_{mkkm})$$

Using Eq. (7.1.29a) and changing the summation index from k to n in the second term gives

▶ $$\dot{\rho}(t)_{mm} = \sum_{n \neq m} \rho(t)_{nn} W_{mn} - \rho(t)_{mm} \sum_{n \neq m} W_{nm} \qquad (7.2.1)$$

Equation (7.2.1) can be interpreted as follows. The diagonal element $\rho(t)_{mm}$ gives the probability of finding the atomic level $|m\rangle$ occupied at time t. This probability increases in time owing to transitions from all other levels $|n\rangle$ to $|m\rangle$. It decreases as a result of transitions from $|m\rangle$ to all other states $|n\rangle$. Thus the rate of change of the diagonal elements must be given in general by a relation of the form

$$\dot{\rho}(t)_{mm} = \text{gain in } |m\rangle - \text{loss from } |m\rangle$$

The "gain" factor is obtained by multiplying $\rho(t)_{nn}$ by the corresponding transition rate $W(n \to m)$ for the transition $|n\rangle \to |m\rangle$ summed over all states $|n\rangle$. The "loss" factor is obtained by multiplying $\rho(t)_{mm}$ by the transition rate $W(m \to n)$ and summing over all n. Thus *the parameters W_{mn} in Eq. (7.2.1) can be interpreted as the probability per unit time that a transition between atomic levels $|n\rangle \to |m\rangle$ is induced by the interaction with the reservoir.*

Equation (7.2.1) is often called the *Pauli Master equation*. The conditions under which this equation hold have been specified in the preceding section. In particular, in order to be able to apply the Markoff approximation it is necessary that the probability of a transition occurring at a given time t depend only on the state of the system at that time and not on its previous history. Equation (7.2.1) plays a prominent role in modern statistics and has been applied to many problems in physics, chemical kinetics, and biology (see, for example, Haken, 1978).

It is instructive to consider the transition rates $W_{mn} = \Gamma^+_{nmmn} + \Gamma^-_{nmmn}$ in more detail by inspecting Eq. (7.1.24). Using Eq. (7.1.13a) and expressing tr_R in terms of the eigenstates $|N\rangle$ of the reservoir Hamiltonian H_R it is found that

$$\int_0^\infty dt'' \exp\left(-i\omega_{mn}t''\right) \text{tr}_R \left[F(t'')_i F_j \rho(0)_R\right]$$

$$= \sum_{N'N} \langle N'|F_i|N\rangle\langle N|F_j|N'\rangle\langle N'|\rho(0)_R|N'\rangle$$

$$\times \int_0^\infty dt'' \exp\left[i(E_{N'} - E_N - \hbar\omega_{mn})t''\right]/\hbar) \qquad (7.2.2a)$$

and the integral occurring in the quantity Γ^-_{nmmn} is given by

$$\int_0^\infty dt'' \exp\left(i\omega_{mn}t''\right) \text{tr}_R \left[F_j F(t'')_i \rho(0)_R\right]$$

$$= \sum_{N'N} \langle N'|F_j|N\rangle\langle N|F_i|N'\rangle\langle N'|\rho(0)_R|N'\rangle$$

$$\times \int_0^\infty dt'' \exp\left[i(E_{N'} - E_N - \hbar\omega_{mn})t''\right]/\hbar) \qquad (7.2.2b)$$

where t'' has been substituted by $-t''$ in the integral. Using Eq. (7.1.11)

$$\sum_i \langle m|Q_i|n\rangle\langle N'|F_i|N\rangle = \langle mN'|V|nN\rangle \qquad (7.2.2c)$$

and inserting Eqs. (7.2.2) into Eqs. (7.2.24) gives

$$W_{mn} = \Gamma^+_{nmmn} + \Gamma^-_{nmmn}$$

$$= (1/\hbar^2) \sum_{ijNN'} \langle n|Q_i|m\rangle\langle m|Q_j|n\rangle\langle N'|F_i|N\rangle\langle N|F_j|N'\rangle\langle N'|\rho_R|N'\rangle$$

$$\times \int_0^\infty dt'' \exp\left[i(E_{N'} - E_N - \hbar\omega_{mn})t''/\hbar\right]$$

$$+ (1/\hbar^2) \sum_{ijNN'} \langle n|Q_j|m\rangle\langle m|Q_i|n\rangle\langle N'|F_j|N\rangle\langle N|F_i|N'\rangle\langle N'|\rho(0)_R|N'\rangle$$

$$\times \int_0^\infty dt'' \exp\left[i(E_{N'} - E_N - \hbar\omega_{nm})t''/\hbar\right]$$

$$= (1/\hbar^2) \sum_{NN'} \langle nN'|V|mN\rangle\langle mN|V|nN'\rangle$$

$$\times \langle N'|\rho(0)_R|N'\rangle \int_{-\infty}^{+\infty} dt'' \exp\left[i(E_{N'} - E_N - \hbar\omega_{mn})t''/\hbar\right)$$

$$= (2\pi/\hbar) \sum_{NN'} |\langle mN|V|nN'\rangle|^2 \langle N'|\rho(0)_R|N'\rangle \delta(E_{N'} - E_N - \hbar\omega_{mn})$$

$$(7.2.3)$$

In Eq. (7.2.3) the element $|\langle mN|V|nN'\rangle|^2$ is the probability of an atom making a transition from level $|n\rangle$ to the level $|m\rangle$ while the reservoir simultaneously undergoes a transition from a state $|N'\rangle$ with energy $E_{N'}$ to a state $|N\rangle$ with energy E_N where, in order to ensure energy conservation, $E_N - E_{N'} = E_n - E_m$ (see Figure 7.1). These probabilities are then averaged over the thermal distribution of the reservoir in order to obtain the net transition rates W_{mn} for the atomic system. Equation (7.2.3) is known as the "golden rule" for the transition rates.

Since V is Hermitian the transition probabilities $|\langle nN'|V|mN\rangle|^2$ satisfy the condition

$$|\langle nN'|V|mN\rangle|^2 = |\langle mN|V|nN'\rangle|^2 \qquad (7.2.4)$$

that is, a transition $|n\rangle|N'\rangle \to |m\rangle|N\rangle$ has the same probability of occurring as the reverse transition. The condition (7.2.4) in general does not, however, apply to the probabilities W_{mn}, describing the net transition $|n\rangle \to |m\rangle$ averaged over the reservoir states. Since the reservoir remains in thermal equilibrium (as discussed in Section 7.1.1) the reservoir is more likely to be in the lower state $|N'\rangle$ of Figure 7.1 than in the upper state $|N\rangle$. Hence if $E_n > E_m$, a transition from an atomic level $|n\rangle$ to a level $|m\rangle$ is more probable than the inverse transition and, in general,

$$W_{mn} \neq W_{nm} \qquad (7.2.5)$$

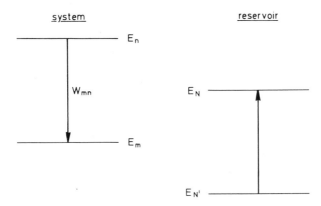

Figure 7.1.   See text for explanations.

Let us discuss these results in more detail. From Eq. (2.6.4)

$$\langle N'|\rho(0)_R|N'\rangle = \exp(-\beta E_{N'})/Z \qquad (7.2.6)$$

which gives

$$W_{mn} = (2\pi/\hbar Z) \sum_{NN'} |\langle mN|V|nN'\rangle|^2 \exp(-\beta E_{N'})\delta(E_{N'} - E_N - \hbar\omega_{mn})$$

$$(7.2.7a)$$

and for the reverse transition

$$W_{nm} = (2\pi/\hbar Z) \sum_{NN'} |\langle nN'|V|mN\rangle|^2 \exp(-\beta E_N)\delta(E_N - E_{N'} - \hbar\omega_{nm})$$

$$(7.2.7b)$$

Using the symmetry property (7.2.4) and energy conservation $E_N = E_{N'} + E_n - E_m$, Eq. (7.2.7b) can be written in the form

$$W_{nm} = \exp[-\beta(E_n - E_m)](2\pi/\hbar Z)$$

$$\times \sum_{N'N} |\langle mN|V|nN'\rangle|^2 \exp(-\beta E_{N'})\delta(E_{N'} - E_N - \hbar\omega_{mn})$$

and from Eq. (7.2.7a) it follows then that

$$\frac{W_{mn}}{W_{nm}} = \frac{\exp(-\beta E_m)}{\exp(-\beta E_n)} \qquad (7.2.8)$$

Hence if $E_n > E_m$ a transition from an atomic level $|n\rangle$ to a level $|m\rangle$ is more probable than the inverse transition.

For example, consider a two-level system, an atomic ground state $|1\rangle$ with energy E_1 and an excited state $|2\rangle$ with energy E_2. From Eqs. (7.2.1) and

(7.2.8) we obtain

$$\dot{\rho}(t)_{11} = W_{12}\rho(t)_{22} - W_{21}\rho(t)_{11}$$

$$= W_{21}\{\exp\left[-\beta(E_1 - E_2)\right]\rho(t)_{22} - \rho(t)_{11}\}$$

$$= -\dot{\rho}(t)_{22} \qquad (7.2.9)$$

Equilibrium is established when the net population of the two levels is constant, that is, when $\dot{\rho}_{11} = \dot{\rho}_{22} = 0$. In this case it follows from Eq. (7.2.9) that the population probabilities are given by a Boltzmann distribution

$$\frac{\rho_{11}}{\rho_{22}} = \frac{\exp\left(-\beta E_1\right)}{\exp\left(-\beta E_2\right)} \qquad (7.2.10)$$

That is, when the initial distribution differs from Eq. (7.2.10) the transitions caused by relaxation processes tend to produce the thermal equilibrium distribution (7.2.10) in which the system is more likely in the lower state $|1\rangle$ than in the upper state $|2\rangle$.

Finally, we note that the result (7.2.5) follows formally from the fact that the reservoir operators F_i and F_j in Eq. (7.1.24) do not in general commute. [Otherwise, by interchanging F_i and F_j and m and n in Eq. (7.1.24) it follows that $\Gamma^{\pm}_{nmmn} = \Gamma^{\pm}_{mnnm}$ and $W_{mn} = W_{nm}$.] On the other hand, in theories in which the reservoir is treated classically and its effect on the system described in terms of random functions of the time rather than in terms of noncommuting operators, it follows that $W_{mn} = W_{nm}$. This is a serious shortcoming of all semiclassical theories of relaxation. A further discussion of this point can be found, for example, in Abragam (1961).

7.3. Kinetics of Stimulated Emission and Absorption

In this section and Section 7.4 we will discuss the physical importance and the application of the basic equations (7.1.28), (7.1.33), and (7.2.1). We will consider the interactions between atoms or molecules with external electromagnetic fields in the presence of relaxation processes. This kind of problem is particularly important in the analysis of quantum electronic devices.

We will confine ourselves to cases in which the energy differences between the atomic or molecular states are sufficiently small for the frequency ν, associated with the transition, to lie in the Hertzian range (that is, $\nu \gtrsim 10^9 - 10^{12}$ Hz corresponding to wavelengths $\lambda \gtrsim 1$ mm). This range includes in particular *radiofrequencies* ($10^4 - 10^9$ Hz) and *microwaves* ($10^9 - 10^{12}$ Hz). The principal spectral transitions in this range are rotational lines of molecules in the millimeter and centimeter range, electronic paramagnetic resonance, and nuclear magnetic resonance. The atomic transitions studied correspond to Zeeman levels produced by external magnetic fields or to natural fine- and hyperfine-structure states. In the following we will use the abbreviation rf to denote frequencies in this range.

One of the special features of transitions in the microwave and radio-frequency region is the *predominance of stimulated emission*. It follows from Einstein's theory of radiation that the ratio between induced and spontaneous transitions is proportional to $\rho(\nu)\lambda^3$ where $\rho(\nu)$ is the spectral radiation density. In the optical region $\rho(\nu)$ and λ are small and—with the exception of the special case of lasers—spontaneous emission dominates. In the rf range λ is large and waves can be produced with large $\rho(\nu)$, so that stimulated emission dominates and spontaneous emission can often be neglected. A further difference between optical and rf lines is that the width of optical lines from conventional light sources is determined by the Doppler effect. For rf transitions the Doppler effect is small and compared to other broadening effects can often be neglected as will be discussed below.

We will now illustrate the application of the basic equations, derived in Sections 7.1 and 7.2, with some simple examples using two-level systems, an atomic or molecular ground state $|1\rangle$ of energy E_1, and an excited state $|2\rangle$ of energy E_2. In magnetic resonance experiments the energy difference between the two spin states is produced by a homogeneous static field H_0. The atomic system is axially symmetric with respect to the quantization axis, which is defined by the direction of H_0 and hence there is no coherence between the two levels. The corresponding density matrix is therefore diagonal in the representation in which the basis states are $|1\rangle$ and $|2\rangle$. In thermal equilibrium the population of the two states is determined by a Boltzmann distribution.

If a rf field is applied it will produce transitions between the states. We will assume that the field is perpendicular to the quantization axis so that there is a privileged *transverse* direction. In this case a coherent super-position of the states $|1\rangle$ and $|2\rangle$ will be produced as discussed in Chapters 4 and 6 and, hence, the reduced density matrix $\rho(t)_S$ of the atomic system of interest will no longer be diagonal.

In addition to the interaction with the external field relaxation processes must be taken into account. The various random interactions between neighboring atoms tend to establish or maintain thermal equilibrium in the medium, that is, a distribution of atoms amongst the two levels conforming to Boltzmann's law. In vapor these interactions occur in collisions between the atoms of the vapor and the walls of the container. In magnetic resonance problems fluctuating fields are produced by the magnetic dipoles of the atoms. In solids there are always interactions between neighboring atoms which are vibrating about their equilibrium positions constituting a reserve of energy. In the following it will always be assumed that the basic approximations of Section 7.1 apply. In particular it will be assumed that the "surroundings" of the atoms under consideration can always be considered as a heat reservoir at thermal equilibrium.

It is therefore necessary to consider two competing processes. The relaxation tends to reestablish thermal distribution of atoms between the two levels. Since the transition probabilities for stimulated emission and absorption are equal the external field tries to equalize the populations (in the absence of spontaneous emission). Eventually, a dynamic equilibrium is established resulting from the competition between these processes in which the population of level $|2\rangle$ is higher than in thermal equilibrium. The relaxation causes more transitions $|2\rangle \rightarrow |1\rangle$ than in the reverse direction and the rf field continually produces more transitions from the lower to the higher level than in the opposite direction. There are therefore more transitions in which photons are absorbed than transitions involving stimulated emission. Energy is therefore continually transferred from the field to the atomic system, while the latter continually restores it to the "reservoir" in form of heat. This absorption of radiation can then be measured using rf spectroscopy.

When an atomic system is under the influence of an external electromagnetic field then, if relaxation is neglected, its Hamiltonian can be written as

$$H(t) = H_0 + V(t) \tag{7.3.1}$$

where H_0 is the Hamiltonian in the absence of the alternating field (in magnetic resonance problems the static field H_0 will be included in H_0). The interaction between the atoms or molecules with the applied field will be represented by

$$V(t) = V \cos \omega t = (1/2)V[\exp(i\omega t) + \exp(-i\omega t)] \tag{7.3.2}$$

In the case of electric dipole transitions, for example, the interaction operator is given by

$$V(t) = -e\mathbf{r}\mathbf{E}(t) = -e\mathbf{r}\mathbf{E} \cos \omega t \tag{7.3.3a}$$

where $e\mathbf{r}$ is the atomic dipole operator and $\mathbf{E}(t)$ the electric field strength. In the case of the interaction between paramagnetic atoms or ions with an alternating electromagnetic field having a magnetic field vector $\mathbf{H}(t)$ is

$$V(t) = -\boldsymbol{\mu}\mathbf{H}(t) = -\boldsymbol{\mu}\mathbf{H} \cos \omega t \tag{7.3.3b}$$

where $\boldsymbol{\mu}$ is the magnetic dipole moment of the atoms (see Section 2.5). It should be noted that the time dependence assumed in Eq. (7.3.2) ensures the

Hermiticity of the operators (7.3.3). For transverse fields the matrix elements $\langle m'|V|m\rangle$ are only nonzero if $m' \neq m$ ($m', m = 1, 2$).

In the absence of any relaxation process the equation of motion for the reduced density matrix of interest is given by the Liouville equation (2.4.20), which we now write in the form

$$[\dot{\rho}(t)_{m'm}] = -i\omega_{m'm}\rho(t)_{m'm} - (i/\hbar)\langle m'|[V(t), \rho(t)_s]|m\rangle \quad (7.3.4a)$$

where $\omega_{m'm} = (E_{m'} - E_m)/\hbar$. The interaction between the atoms and their surroundings is usually taken into account by adding the relevant relaxation term

$$\rho(t)_{m'm} = \sum_{nn'} R_{m'mn'n}\rho(t)_{n'n} \quad (7.3.4b)$$

to the expression (7.3.4a) to give the total equation of motion:

▶ $$\dot{\rho}(t)_{m'm} = -i\omega_{m'm}\rho(t)_{m'm} - (i/\hbar)\langle m'|[V(t), \rho(t)_s]|m\rangle$$
$$+ \sum_{nn'} R_{m'mn'n}\rho(t)_{n'n} \quad (7.3.4)$$

which is the relevant Master equation. *It should be noted that this relation assumes that the coupling between the various terms in the time evolution of $\rho(t)_s$ can be ignored.*

In the optical region an additional term $(\dot{\rho}_{m'm})_{sp}$ must be added to Eq. (7.3.4) to represent the effect of spontaneous decay. Since this is essentially a random process, which is caused by the fluctuations of the vacuum field, spontaneous emission can be represented by

$$(\dot{\rho}_{m'm})_{sp} = -\Gamma_{m'm}\rho_{m'm}$$

where $\Gamma_{m'm}$ is the spontaneous decay rate of the excited level.

Using Eq. (7.2.9) the equations for the diagonal elements can be written in the form

$$\dot{\rho}(t)_{11} = -(i/\hbar)\langle 1|[V(t), \rho(t)_s]|1\rangle + W_{12}\rho(t)_{22} - W_{21}\rho(t)_{11} \quad (7.3.5a)$$

$$\dot{\rho}(t)_{22} = -(i/\hbar)\langle 2|[V(t), \rho(t)_s]|2\rangle + W_{21}\rho(t)_{11} - W_{12}\rho(t)_{22} \quad (7.3.5b)$$

In these equations the first term

$$(\dot{\rho}_{mm})_{rad} \equiv -(i/\hbar)\langle m|[V(t), \rho(t)_s]|m\rangle \quad (7.3.6)$$

is the rate of change of the population probability of the level $|m\rangle$ induced by the rf field and the other terms describe the influence of the relaxation processes.

It follows from Eq. (7.1.33) that the off-diagonal terms are described by

$$\dot{\rho}(t)_{21} = \dot{\rho}(t)_{12}^*$$

$$= -i(\omega_{21} - i\gamma_{21})\rho(t)_{21} - (i/\hbar)\langle 2|[V(t), \rho(t)_s]|1\rangle$$

$$= -i(\omega_{21} - i\gamma_{21})\rho(t)_{21}$$

$$-(i/2\hbar)\langle 2|V|1\rangle[\exp(i\omega t) + \exp(-i\omega t)](\rho_{11} - \rho_{22}) \quad (7.3.7)$$

where Eq. (7.3.2) has been applied together with the fact that the diagonal elements of V vanish.

Equations (7.3.5) and (7.3.7) give the rate of change of the density matrix elements under the combined influence of the external field and relaxation. Dynamic equilibrium is established when $\dot{\rho}_{11} = \dot{\rho}_{22} = 0$, that is, when the effects of the stimulated emission and absorption are balanced by the relaxation processes. We will now study this "stationary" solution in some detail.

Let us first consider Eq. (7.3.7). In the interaction representation the elements $\rho_I(t)_{m'm}$ are related to $\rho(t)_{m'm}$ by

$$\rho(t)_{21} = \exp(-i\omega_{21}t)\rho_I(t)_{21} \quad (7.3.8)$$

where Eq. (2.4.37) has been used. Inserting this relation into Eq. (7.3.7) and multiplying both sides with $\exp(i\omega_{21}t)$ we obtain

$$\dot{\rho}_I(t)_{21} = -\gamma_{21}\rho_I(t)_{21} - (i/2\hbar)\langle 2|V|1\rangle\{\exp[i(\omega_{21} + \omega)t]$$

$$+ \exp[i(\omega_{21} - \omega)t]\}[\rho(t)_{11} - \rho(t)_{22}] \quad (7.3.9a)$$

In the resonance region $\omega_{21} \approx \omega$ the dominant contribution will come from the low-frequency term $\exp[i(\omega_{21} - \omega)t]$ and, in first-order approximation, the rapidly varying terms $\exp[i(\omega_{21} + \omega)t] \approx \exp(2i\omega t)$ can be neglected (this is referred to as the "rotating wave approximation"). Equation (7.3.9a) then simplifies to

$$\dot{\rho}_I(t)_{21} = -\gamma_{21}\rho_I(t)_{21} - (i/2\hbar)\langle 2|V|1\rangle \exp[i(\omega_{21} - \omega)t][\rho(t)_{11} - \rho(t)_{22}]$$

$$(7.3.9)$$

In order for the system to be in a steady state the density matrix elements must be independent of the time at which they are calculated. Since the "driving" term in Eq. (7.3.9) varies as $\exp[i(\omega_{21} - \omega)t]$ we seek a solution of the form

$$\rho_I(t)_{21} = \exp[i(\omega_{21} - \omega)t]\rho(\omega)_{21} \quad (7.3.10a)$$

corresponding to

$$\rho(t)_{21} = \exp(-i\omega t)\rho(\omega)_{21} \quad (7.3.10b)$$

in the Schrödinger picture. When inserting Eq. (7.3.10a) into the expression (7.3.9) the time-dependent exponential factors cancel and

$$\rho(\omega)_{21} = -\frac{i}{2\hbar}\langle 2|V|1\rangle \frac{\rho_{11} - \rho_{22}}{i(\omega_{21} - \omega) + \gamma_{21}}$$

$$= \frac{1}{2\hbar}\langle 2|V|1\rangle \frac{\rho_{11} - \rho_{22}}{\omega - \omega_{21} + i\gamma_{21}} \qquad (7.3.11a)$$

The elements $\rho(t)_{12}$ can be determined in a similar way. It should be noted that in this case the term $\sim\exp(i\omega t)$ gives the dominant contribution in the rotated wave approximation. Hence

$$\rho(t)_{12} = \exp(i\omega t)\rho(\omega)_{12} \qquad (7.3.11b)$$

and

$$\rho(\omega)_{12} = \frac{1}{2\hbar}\langle 2|V|1\rangle^* \frac{\rho_{11} - \rho_{22}}{\omega - \omega_{21} - i\gamma_{21}^*} \qquad (7.3.11c)$$

where the relation (7.1.32) has been applied.

It follows from the above discussion that the rotating wave approximation is obtained by setting

$$\langle 2|V(t)|1\rangle = (1/2)\langle 2|V|1\rangle \exp(-i\omega t)$$
$$\langle 1|V(t)|2\rangle = (1/2)\langle 1|V|2\rangle \exp(i\omega t) \qquad (7.3.12)$$

for the interaction (7.3.2) and using Eqs. (7.3.10b) and (7.3.11b) for the density matrix elements. The higher-order terms $\sim \exp(2i\omega t)$ neglected here are the basic source of nonlinear effects in quantum electronics.

Consider now Eqs. (7.3.5) which describe the diagonal density matrix elements. Equation (7.3.6) can be rewritten as follows: Applying again the rotating wave approximation, using Eqs. (7.3.10b), (7.3.11b), and (7.3.12), and substituting Eq. (7.3.11) for the off-diagonal elements we obtain

$$[\dot{\rho}(t)_{22}]_{\mathrm{rad}} = -\frac{i}{\hbar}\langle 2|V(t)|1\rangle\rho(t)_{12} + \frac{i}{\hbar}\langle 1|V(t)|2\rangle\rho(t)_{21}$$

$$= -\frac{i}{2\hbar^2}|\langle 2|V|1\rangle|^2 (\rho_{11} - \rho_{22})$$

$$\times \left(\frac{1}{\omega - \omega_{21} - i\gamma_{21}^*} - \frac{1}{\omega - \omega_{21} + i\gamma_{21}}\right)$$

$$= \frac{1}{\hbar^2}|\langle 2|V|1\rangle|^2 \frac{\gamma_{21}'}{(\omega - \omega_{21} - \gamma_{21}'')^2 + \gamma_{21}'^2}(\rho_{11} - \rho_{22}) \qquad (7.3.13a)$$

where γ'_{21} (γ''_{21}) denotes the real (imaginary) part of the relaxation parameter γ_{21}.
  Similarly,

$$[\dot{\rho}(t)_{11}]_{\text{rad}} = \frac{1}{\hbar^2}|\langle 2|V|1\rangle|^2 \frac{\gamma'_{21}}{(\omega - \omega_{21} - \gamma''_{21})^2 + \gamma'^2_{21}}(\rho_{22} - \rho_{11}) \quad (7.3.13b)$$

Equations (7.3.13) describe the change of the population probability of the two levels induced by the rf field. Comparing both expressions it can be seen that

$$[\dot{\rho}(t)_{11}]_{\text{rad}} = -[\dot{\rho}(t)_{22}]_{\text{rad}} \quad (7.3.13)$$

This result can be readily understood by noting that the field can increase the number of atoms in level $|1\rangle$ only by inducing transitions $|2\rangle \rightarrow |1\rangle$ and, vice versa, a "gain" in the population of level $|2\rangle$ is due to induced transitions $|1\rangle \rightarrow |2\rangle$.
  Defining

$$W(\omega)_{21} = \frac{1}{\hbar^2}|\langle 2|V|1\rangle|^2 \frac{\gamma'_{21}}{(\omega - \omega_{21} - \gamma''_{21})^2 + \gamma'^2_{21}}$$

$$= W(\omega)_{12} \quad (7.3.14)$$

then, in the steady state case, Eqs. (7.3.5) can be written in the form

$$\dot{\rho}(t)_{11} = [W_{12} + W(\omega)_{12}]\rho_{22} - [W_{21} + W(\omega)_{21}]\rho_{11} = 0$$
$$\dot{\rho}(t)_{22} = [W_{21} + W(\omega)_{21}]\rho_{11} - [W_{12} + W(\omega)_{12}]\rho_{22} = 0 \quad (7.3.15)$$

In Section 7.2 it was shown that the parameters W_{12} and W_{21} are the probabilities for the transitions $|2\rangle \rightarrow |1\rangle$ and $|1\rangle \rightarrow |2\rangle$, respectively, caused by the relaxation mechanism. Similarly, the parameters $W(\omega)_{12}$ and $W(\omega)_{21}$ are the probabilities for a transition $|2\rangle \rightarrow |1\rangle$ and $|1\rangle \rightarrow |2\rangle$ induced by an alternating field with frequency ω. We can therefore consider $\rho(t)_{22}[W_{12} + W(\omega)_{12}]$ and $\rho(t)_{11}[W_{21} + W(\omega)_{21}]$ in Eqs. (7.3.15) as the rates at which the population probability of level $|1\rangle$ increases and decreases under the combined influence of the external field and the relaxation processes.
  If the intensity of the rf field is sufficiently high then the population probabilities $\rho(t)_{11}$ and $\rho(t)_{22}$ can be considerably different from their values $\rho_{11}^{(0)}$ and $\rho_{22}^{(0)}$ at thermal equilibrium. In this case the atoms are said to be *pumped* by the field. If the intensity of the field is weak then the diagonal elements remain close to their values at thermal equilibrium and on the right-hand sides of Eqs. (7.3.11a) and (7.3.11c)

$$\rho_{mm} \approx \rho_{mm}^{(0)} \quad (7.3.16)$$

The power absorbed per unit time from the field by the atoms is

$$\frac{dE}{dt} = E_1[\dot{\rho}(t)_{11}]_{\text{rad}} + E_2[\dot{\rho}(t)_{22}]_{\text{rad}}$$

$$= (E_1 - E_2)[\dot{\rho}(t)_{11}]_{\text{rad}} \tag{7.3.17}$$

where the relaxation (7.3.13) has been applied. If $[\dot{\rho}(t)_{11}]_{\text{rad}} < 0$ the induced transitions $|1\rangle \rightarrow |2\rangle$ exceed the number of downward transitions $|2\rangle \rightarrow |1\rangle$ and the corresponding energy is absorbed from the rf field. Since $E_2 > E_1$ then in this case $dE/dt > 0$. Conversely, if $[\dot{\rho}(t)_{11}]_{\text{rad}} < 0$ more energy is given out by the stimulated emission process than is absorbed and $dE/dt < 0$.

Substitution of Eq. (7.3.13a) into Eq. (7.3.17) yields

$$\frac{dE}{dt} = \frac{1}{\hbar^2}(E_2 - E_1)|\langle 2|V|1\rangle|^2 \frac{\gamma'_{21}}{(\omega - \omega_{21} - \gamma''_{21})^2 + \gamma'^2_{21}}(\rho_{11} - \rho_{22}) \tag{7.3.18a}$$

and for weak fields

$$\frac{dE}{dt} = \frac{1}{\hbar^2}(E_2 - E_1)|\langle 2|V|1\rangle|^2 \frac{\gamma'_{21}}{(\omega - \omega_{21} - \gamma''_{21})^2 + \gamma'^2_{21}}(\rho_{11}^{(0)} - \rho_{22}^{(0)}) \tag{7.3.18b}$$

Since $\rho_{11}^{(0)} > \rho_{22}^{(0)}$ at thermal equilibrium it follows from Eq. (7.3.18b) that $dE/dt > 0$ and energy is absorbed from the field. In this case Eq. (7.3.18b) shows that the presence of relaxation has two effects on the absorption line: (i) a *line shift* caused by the imaginary part γ''_{21} and (ii) a *line broadening* due to the real part γ'_{21} of the parameter γ_{21}.

If, in any circumstances, $\rho(t)_{11} < \rho(t)_{22}$, which in the case of lasers and masers is called *population inversion*, then $dE/dt < 0$. This means that on passing through the medium radiation is not attenuated by absorption but *amplified* by the stimulated emission. The operation of lasers and masers is based on this effect.

7.4. The Bloch Equations

7.4.1. Magnetic Resonance

In this section we will apply Eq. (7.3.4) to magnetic resonance problems. The simplest possible system in which a magnetic resonance can be observed is a two-level system, for example, atoms or molecules with orbital angular momentum zero and spin $1/2$, or atoms with no electronic angular momentum and nuclear spin $1/2$.

If a static external magnetic field $\mathbf{H}_0$ is applied in the z direction then the energies of the two spin states $|1\rangle$ and $|2\rangle$, corresponding to "spin up" and "spin down," are given by

$$E_1 = -\mu|\mathbf{H}_0|, \qquad E_2 = \mu|\mathbf{H}_0| \qquad (7.4.1)$$

(see Figure 7.2, where it has been assumed that the magnetic moment μ is positive). The energy splitting is

$$\Delta E = E_2 - E_1 = 2\mu|\mathbf{H}_0| \qquad (7.4.2)$$

It follows that if a transverse electromagnetic field with magnetic field strength $\mathbf{H}_1(t)$ oscillating at an angular frequency ω_L and satisfying the resonance condition

$$\hbar\omega_L = \Delta E \qquad (7.4.3)$$

(Larmor frequency) is applied to the system absorption of this energy will occur as electrons (or nuclei) are excited from the lower to the upper level. The required frequencies are well into the microwave region in the case of paramagnetic or electron spin resonance. Radiofrequencies are required in the case of nuclear magnetic resonance.

The number of atoms in the upper state is increased by the incoming radiation but this increase is reduced through relaxation effects which transfer energy from the excited state to the "surroundings" and attempt to reestablish the conditions of thermal equilibrium. The effect of these competing factors on the spin density matrix $\rho(t)_S$ is described by Eq. (7.3.4) where now H_0 includes the static field $\mathbf{H}_0$:

$$H_0 = H_0' - \boldsymbol{\mu} \cdot \mathbf{H}_0 \qquad (7.4.4)$$

The Hamiltonian H_0' of the unperturbed spin states does not appear in the equation of motion because of the degeneracy of these states. Using Eq. (7.3.3b) the interaction between the spin and the external fields is represented by the Hamiltonian

$$H(t) = -\boldsymbol{\mu} \cdot \mathbf{H}_0 + V(t) = -\boldsymbol{\mu} \cdot (\mathbf{H}_0 + \mathbf{H}(t)_1) \qquad (7.4.5)$$

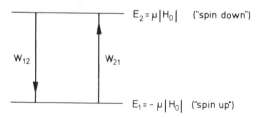

Figure 7.2. Energy level splitting in a static magnetic field.

First of all we will consider Eqs. (7.3.5) in the absence of the rf field. In this case

$$\dot{\rho}(t)_{11} = W_{12}\rho(t)_{22} - W_{21}\rho(t)_{11} = -\dot{\rho}(t)_{22} \qquad (7.4.6)$$

It will be assumed that each spin under consideration reacts to the external fields independently of all other spins and that the "surroundings" can be considered as a heat reservoir at thermal equilibrium. Adding and subtracting the term $W_{12}\rho_{11}$ to Eq. (7.4.6) then, since $\rho_{11} + \rho_{22} = 1$, it follows that

$$\dot{\rho}(t)_{11} = W_{12} - (W_{12} + W_{21})\rho(t)_{11} \qquad (7.4.7a)$$

and

$$\dot{\rho}(t)_{22} = W_{21} - (W_{12} + W_{21})\rho(t)_{22} \qquad (7.4.7b)$$

which gives

$$\dot{\rho}(t)_{11} - \dot{\rho}(t)_{22} = (W_{12} - W_{21}) - (W_{12} + W_{21})[\rho(t)_{11} - \rho(t)_{22}] \quad (7.4.8)$$

In thermal equilibrium $\dot{\rho}_{11} = \dot{\rho}_{22} = 0$ and it follows from Eq. (7.4.8) that

$$\rho_{11}^{(0)} - \rho_{22}^{(0)} = \frac{W_{12} - W_{21}}{W_{12} + W_{21}} \qquad (7.4.9)$$

where $\rho_{11}^{(0)}$ and $\rho_{22}^{(0)}$ denote the population probability of levels $|1\rangle$ and $|2\rangle$, respectively, at thermal equilibrium (in the presence of the static field). Defining the parameter T_1 by

▶ $$T_1 = 1/(W_{12} + W_{21}) \qquad (7.4.10)$$

Eq. (7.4.8) can be written as

$$\dot{\rho}(t)_{11} - \dot{\rho}(t)_{22} = \frac{(\rho_{11}^{(0)} - \rho_{22}^{(0)}) - [\rho(t)_{11} - \rho(t)_{22}]}{T_1} \qquad (7.4.11)$$

Note that T_1 is real (because of Eq. (7.1.30)].

The rf field can be taken into account by adding the relevant terms to Eq. (7.4.11):

$$\dot{\rho}(t)_{11} - \dot{\rho}(t)_{22} = -\frac{i}{\hbar}[\langle 1|[V(t), \rho(t)_S]|1\rangle - \langle 2|[V(t), \rho(t)_S]|2\rangle]$$

$$+ \frac{(\rho_{11}^{0} - \rho_{22}^{0}) - [\rho(t)_{11} - \rho(t)_{22}]}{T_1} \qquad (7.4.12)$$

We will discuss the off-diagonal elements in an approximation where the imaginary part of γ_{12} (that is, the line shift) is neglected. In this case, defining

▶ $$T_2 = \frac{1}{\gamma_{12}} \qquad (7.4.13)$$

the off-diagonal elements obey the equation

$$\dot{\rho}(t)_{21} = (-i\omega_{12} + 1/T_2)\rho(t)_{21} - (i/\hbar)\langle2|[V(t), \rho(t)_S]|1\rangle$$
$$= \dot{\rho}(t)_{12}^*$$
(7.4.14)

which can be rewritten as

$$\dot{\rho}(t)_{21} = -\frac{i}{\hbar}\langle2|[H(t), \rho(t)_S]|1\rangle + \frac{\rho(t)_{21}}{T_2}$$
(7.4.15)

where $H(t)$ is the Hamiltonian (7.4.5).

The macroscopic magnetization **M** is given by Eq. (2.6.6),

$$M_i = N\gamma\hbar\langle\sigma_i\rangle/2$$
(7.4.16a)

$(i = x, y, z)$, where σ_i is the corresponding Pauli matrix, N the total number of atoms per unit volume, and γ the gyromagnetic ratio. Using the expressions for the Pauli matrices [Eq. (1.1.6)] **M** is given explicitly by

$$M_x = \tfrac{1}{2}N\gamma\hbar(\rho_{12} + \rho_{21}), \qquad M_y = \tfrac{1}{2}N\gamma\hbar i(\rho_{12} - \rho_{21}),$$

$$M_z = \frac{N\gamma\hbar}{2}(\rho_{11} - \rho_{22})$$
(7.4.16b)

In the absence of relaxation effects the equation of motion for the magnetization **M** is given by Eq. (2.5.5) with **P** replaced by **M** and **H** replaced by $\mathbf{H}_0 + \mathbf{H}_1(t)$. Adding the relaxation terms and using Eqs. (7.4.11), (7.4.15), and (7.4.16b) we obtain

▶
$$\frac{dM_x}{dt} = \gamma[\mathbf{M} \times (\mathbf{H}_0 + \mathbf{H}(t)_1]_x - \frac{M_x}{T_2}$$
(7.4.17a)

$$\frac{dM_y}{dt} = \gamma[\mathbf{M} \times (\mathbf{H}_0 + \mathbf{H}(t)_1)]_y - \frac{M_y}{T_2}$$
(7.4.17b)

$$\frac{dM_z}{dt} = \gamma[\mathbf{M} \times \mathbf{H}(t)_1]_z + \frac{M_z^{(0)} - M_z}{T_1}$$
(7.4.17c)

where

$$M_z^{(0)} = \frac{W_{12} - W_{21}}{W_{12} + W_{21}}$$
(7.4.18)

As shown by Eq. (7.4.17c) $M_z^{(0)}$ is the value of M_z at thermal equilibrium $(dM_z/dt = 0)$ in the absence of the rf field.

The set of equations (7.4.17) are the *Bloch equations* and were first derived by Bloch (1946) for N-level atoms and further developed by Bloch and Wangsness (1952). The main feature of the general Bloch equations is that the effect of the relaxation processes is described in terms of two real parameters, T_1 and T_2.

The principal aim of the discussion given above is to elucidate the various approximations which are incorporated in the Bloch equations. These approximations are not always valid and relaxation is not in general as simple as expressed by the equations (7.4.17). Nevertheless, these equations describe the observed phenomena to a good approximation in a large number of cases. It should be noted that the macroscopic magnetization $\mathbf{M}$ obeys the same equation as in classical phenomenological theories and it can be shown that the values of $\mathbf{M}$ are equal to the values calculated in classical models. This is a consequence of the fact that spontaneous emission is neglected in Eq. (7.3.4). A detailed discussion of this point can be found in Abragam (1961). Finally, we note that $W_{12} = W_{21}$ in semiclassical theories (as discussed in Section 7.2), from which follows $M_z^{(0)} = 0$, which does not agree with experiment.

7.4.2. Longitudinal and Transverse Relaxation. Spin Echoes

A detailed discussion of the Bloch equations can be found in many treatments of magnetic resonance phenomena and we will confine ourselves to a discussion of the physical nature of the parameters T_1 and T_2.

Let us assume that at a certain time (say, at $t = 0$) the field $\mathbf{H}(t)_1$ is removed. In the absence of $\mathbf{H}_1$ the Bloch equations reduce to

$$\frac{dM_x}{dt} = \omega_L M_y - \frac{M_y}{T_2}$$

$$\frac{dM_y}{dt} = -\omega_L M_x - \frac{M_y}{T_2} \qquad (7.4.19)$$

$$\frac{dM_z}{dt} = \frac{M_z^{(0)} - M_z}{T_1}$$

with $\omega_L = \gamma |\mathbf{H}_0|$. In the absence of any relaxation process $\mathbf{M}$ would precess freely around the static field $\mathbf{H}_0$ with a frequency ω_L. M_z would remain constant and M_x and M_y rotate with constant magnitude in the x–y plane. Because of the various interactions between the spins and their surroundings the spin system will relax to thermal equilibrium. It can be shown that a solution of Eq. (7.4.19) is given by

$$M(t)_x = A \sin (\omega_L t + \varphi) \exp (-t/T_2)$$

$$M(t)_y = A \cos (\omega_L t + \varphi) \exp (-t/T_2) \qquad (7.4.20)$$

$$M(t)_z = (B - M_z^{(0)}) \exp (-t/T_1) + M_z^{(0)}$$

where A, B, and φ are integration constants. Equations (7.3.20) show that, as a consequence of relaxation, M_x and M_y will vanish with a time constant T_2 and M_z will reach its equilibrium value with a time constant T_1. T_2 therefore accounts for the decay of M_x and M_y, which are perpendicular to $\mathbf{H}_0$, and for this reason T_2 is called the *transverse relaxation time*. T_1 accounts for the decay of the longitudinal component M_z of $\mathbf{M}$ and is termed the *longitudinal relaxation time*.

That M_x and M_y vanish at thermal equilibrium is due to the fact that there is no preferred transverse direction. The directions of the individual transverse components therefore vary in a random way from one atom to another and the net resultant is zero. The fact that M_z is different from zero is related to the axial symmetry of the system due to the static field $\mathbf{H}_0$ which produces the difference in energy between the two levels.

The physical nature of T_1 and T_2 can be understood by noting that relaxation is caused by different mechanisms. First of all, the *spin–lattice interaction* includes all processes in which energy is exchanged between the spin system and its surroundings, for example the lattices in a crystal. In general, all degrees of freedom except the spins are called a *lattice*. A transfer of energy from the spin system to the lattice is associated with transitions from the upper to the lower spin state and causes the population number of the two spin states, and hence M_z, to change. *Longitudinal relaxation is therefore associated with an energy transfer from the spin system to the lattice.* T_1 is thus a measure of the time required for the system to reach an energy equilibrium with its environment.

A second type of interaction, known as *spin–spin interaction*, includes all mechanisms whereby the spins can exchange energy among themselves rather than transferring it to the lattice as a whole. For example, in an elastic collision in which one atom undergoes a transition $|1\rangle \rightarrow |2\rangle$ and the other one a transition $|2\rangle \rightarrow |1\rangle$ the energy of the spin system and the value of M_z do not change. Such collisions therefore do not contribute to longitudinal relaxation but destroy the coherence between the spin states (see Chapter 3); the off-diagonal elements of the spin density matrix and the transverse components of $\mathbf{M}$ are therefore reduced. *Transverse relaxation is therefore associated with a loss of coherence of the spin system.* Note that any process contributing to T_1 will in general also destroy the coherence so that $T_1 \geq T_2$. Magnetic resonance techniques allow the determination of the relaxation times and enable information on the various relaxation processes to be obtained (see, for example, Abragam, 1961; Corney, 1977).

The physical meaning of T_2 can be understood in terms of the following simple model (which does not, however, include all the aspects of transverse relaxation). Immediately after the removal of the rf field the individual spins start precessing around $\mathbf{H}_0$. Without relaxation processes all component spins would rotate with the same frequency ω_L and the initial values of the magnitudes M_x and M_y would be constant in time. The various random interactions of the magnetic dipoles produce a magnetic field at each atom giving rise to a fluctuating component in addition to the external field $\mathbf{H}_0$, either aiding or opposing $\mathbf{H}_0$ and causing the individual precession rates to be either faster or slower. As a result the spins get out of step and in the course of time their distribution spreads over a wider and wider range in the x–y plane and the net transverse component eventually vanishes. T_2 is a measure of the time required for the spins to become completely random to each other. A direct measurement of T_2 is the most unambiguous method of investigating the mechanisms by which coherence is lost.

This simple model of transverse relaxation readily explains a phenomenon known as the *spin echo*. Suppose that a set of nuclear magnetic dipoles is such that the magnetization $\mathbf{M}$ of the dipoles points in the direction of the static field (z axis). A resonant radiofrequency field is applied as a pulse of such a duration that it rotates $\mathbf{M}$ from the z to the x direction ("$\pi/2$ pulse"; Figure 7.3a). When the pulse is switched off the individual spins

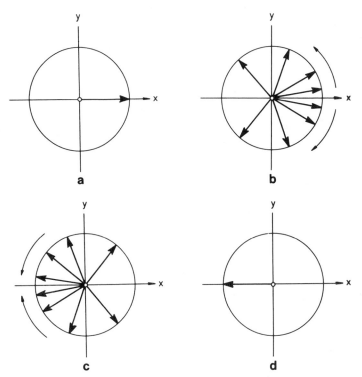

Figure 7.3. Illustration of spin echoes.

precess about the direction of the static field. It is convenient to discuss the motion of the spins in a coordinate system which rotates with the Larmor frequency about the z axis. In the absence of any relaxation process the spins would rotate freely around $\mathbf{H}_0$ with the Larmor frequency, that is, they would appear to be at rest when viewed from the rotating system. Since the precession frequency is slightly different for each of the component moments because of relaxation processes the spins get out of phase and become distributed in the x–y plane (Figure 7.3b). After a time $t < T_2$ a second pulse is applied with such a duration that the direction of all spins are reversed ("π pulse"). That is, the component spins are just turned over as shown by Figure 7.3c. Because the spins keep rotating with their initial speed the spins converge back to one vector (Figure 7.3d), producing a "pulse" of magnetization in this direction, which appears as an "echo" of the first high-frequency pulse.

Abragam (1961) has given the following analog of this phenomenon. Suppose a group of ants is crawling around the edge of a pancake. They all start together in a small area but, because of their different speeds, they will progressively spread out around the circumference ("T_2 process"). If the pancake is turned over ("π pulse"), the ants are turned around but continue to crawl in the same direction. Eventually, all the ants will again end up bunched together with the exception of those which have fallen off the pancake ("T_1 process").

7.4.3. The "Optical" Bloch Equations

As we discussed briefly at the beginning of this section there is a close analogy between a two-level atom and a spin-1/2 system (in a static magnetic field in the z direction). The "spin-down" state corresponds to the atomic ground level and the "spin-up" state to the excited level. The formalism developed above for magnetic resonance phenomena can be generalized to any two-level system driven by a resonant transverse field. This approach can be used to describe experiments in the microwave or optical region when coherent fields (maser, laser) are employed.

The relevance of the Bloch equations to the description of the maser was first recognized by Feynmann, Vernon, and Hellwarth (1957). Following these authors we define a fictitious quantity, the "pseudospin vector" $\mathbf{v}$, with components v_i, analogously to Eqs. (7.4.16b):

$$v_1 = \text{tr}\left[\rho(t)_S\sigma_x\right] = \rho(t)_{12} + \rho(t)_{21}, \qquad v_2 = \text{tr}\left[\rho(t)_S\sigma_y\right] = i[\rho(t)_{12} + \rho(t)_{21}]$$

$$v_3 = \text{tr}\left[\rho(t)_S\sigma_z\right] = \rho(t)_{11} - \rho(t)_{22} \qquad (7.4.21)$$

In these equations $\rho(t)_S$ denotes the density matrix of the atomic two-level system. As shown by Eqs. (7.4.21) v_3 depends on the difference of the population numbers and v_1 and v_2 characterize the coherence between the two states.

Starting with Eqs. (7.3.5) and (7.3.7) it can then be shown, in an analogous way to the derivation of Eqs. (7.4.17), that the components v_i obey the following set of equations, which are termed the *"generalized"* or *"optical" Bloch equations*:

$$\frac{dv_1}{dt} = [\boldsymbol{\omega} \times \mathbf{V}]_1 - \frac{v_1}{T_2}$$

$$\frac{dv_2}{dt} = [\boldsymbol{\omega} \times \mathbf{v}]_2 - \frac{v_2}{T_2} \qquad (7.4.22)$$

$$\frac{dv_3}{dt} = [\boldsymbol{\omega} \times \mathbf{v}]_3 - \frac{v_3}{T_1}$$

where the "vector" $\boldsymbol{\omega}$ has components

$$\omega_1 = (V_{12} + V_{21})/\hbar, \quad \omega_2 = -i(V_{12} - V_{21})/\hbar, \quad \omega_3 = -(E_1 - E_2)/\hbar$$

$$(7.4.23)$$

and $V_{ij} = \langle i|V|j \rangle$ are the matrix elements of the operator V introduced in Section 7.3 which describes the interaction between the atoms and the electromagnetic field. The relaxation times are defined as in Eqs. (7.4.10) and (7.4.13).

The use of Eqs. (7.4.22) enables a geometrical interpretation of electric dipole transitions in a two-level system to be made in a similar way to the description of magnetic resonance phenomena. Without the relaxation terms the equations (7.4.22) can be interpreted as a precession of the "vector" $\mathbf{v}$ about the "vector" $\boldsymbol{\omega}$. The relaxation terms have similar meanings to those in the previous section. The relaxation of v_3 (time constant T_1) is associated with the various ways in which energy can be exchanged between the atoms and their environment. The relaxation of v_1 and v_2 (time constant T_2) corresponds to a loss of coherence caused by the various dephasing processes. This geometrical interpretation can be used to explain photon echoes by considering the precession of the vector $\mathbf{v}$ between two applied pulses. Furthermore, Eqs. (7.4.23) can be used to treat electromagnetic transitions in very strong radiation fields where perturbation methods cannot be applied. It should be noted, however, that, as in the case of the Bloch equations (7.4.17), the equations (7.4.22) do not describe all phenomena associated with transitions in a two-level system. For a detailed discussion of Eqs. (7.4.22) and their applications the reader is referred, for example, to the book by Walther (1976).

7.5. Some Properties of the Relaxation Matrix

7.5.1. General Constraints

In Section 7.1 it was shown that, under the Markoff approximation, the rate of change of the density matrix with respect to time can be represented by a set of coupled linear differential equations, which is in the interaction picture

$$\dot{\rho}_I(t)_{m'm} = \sum_{nn'} R_{m'mn'n}\rho_I(t)_{n'n} \tag{7.5.1}$$

$$[\langle m'|\rho(t)_{SI}|m\rangle \equiv \rho_I(t)_{m'm}]$$

In the secular approximation (7.1.28) we have

$$R_{m'mn'n} = W_{mn}\delta_{n'n} - \gamma_{m'm}\delta_{m'n'}\delta_{mn}$$

where the first term contributes only if $m \neq n$.

The set of the (time-independent) coefficients $R_{m'mn'n}$ is called the *relaxation matrix*. It follows from the discussions in the preceding sections that, for example, R_{mmnn} determines the rate of transfer of atoms from the level $|n\rangle$ to the level $|m\rangle$. Following Happer (1972) we will now show that physical considerations place some important constraints on the elements of the relaxation matrix. In doing this some of the general results obtained in the previous sections will be rederived and generalized.

1. If the relaxation mechanism does not alter the number of atoms present then it follows from the probability interpretation of the diagonal elements that

$$\sum_m \dot{\rho}(t)_{mm} = \sum_{nn'} \left(\sum_m R_{mmn'n}\right)\rho_I(t)_{n'n} = 0$$

Since in general $\rho_I(t)_{n'n} \neq 0$ then

$$\sum_m R_{mmn'n} = 0 \tag{7.5.2}$$

for any n' and n.

2. Consider the equation corresponding to Eq. (7.5.1) for the complex conjugate:

$$\dot{\rho}_I(t)^*_{m'm} = \sum_{nn'} R^*_{m'mn'n}\rho_I(t)^*_{n'n} \tag{7.5.3a}$$

Applying the Hermiticity condition it follows that

$$\dot{\rho}_I(t)_{mm'} = \sum_{nn'} R^*_{m'mn'n}\rho_I(t)_{nn'} \tag{7.5.3b}$$

Equations (7.5.1) and (7.5.3b) imply that

$$\sum_{n'n} \rho_I(t)_{nn'}(R^*_{m'mn'n} - R_{mm'nn'}) = 0$$

which gives

$$R^*_{m'mn'n} = R_{mm'nn'} \tag{7.5.3}$$

3. The diagonal elements of the density matrix are nonnegative and cannot exceed unity (see Section 2.2). In order to ensure that these conditions are met the following conditions must be satisfied:

$$R_{mmmm} \leq 0 \tag{7.5.4a}$$

$$R_{nnmm} \geq 0 \tag{7.5.4b}$$

for any orthogonal set of atomic states $|m\rangle$ and $|n\rangle$. The validity of these conditions can be seen as follows: Suppose that at a certain time t only a single level $|m\rangle$ is populated as described by Eq. (7.5.4a). For any state $n \neq m$ it is then

$$\dot{\rho}(t)_{nn} = R_{nnmm}\rho(t)_{mm}$$

Since $\rho(t)_{nn}$ has its minimum value at time t, then it can only increase or remain zero as described by Eq. (7.5.4b).

7.5.2. Relaxation of State Multipoles

If the system under consideration possesses certain symmetry properties under rotations it is convenient to express the relaxation equation (7.5.1) in terms of state multipoles. For simplicity we will confine ourselves to the case of an atomic system which is in the ground state with sharp angular momentum J, and relaxation processes which cause transitions within this single multiplet.

In a similar way to Eq. (4.3.3) state multipoles can be defined as

$$\langle T(J, t)^\dagger_{KQ}\rangle = \sum_{MM'} (-1)^{J-M'}(2K + 1)\begin{pmatrix} J & J & K \\ M' & -M & -Q \end{pmatrix}\rho_I(t)_{M'M}$$

and using the inverse relaxation (4.3.6) the relaxation equation (7.5.1) can be written as

$$\frac{\partial\langle T(J, t)^\dagger_{KQ}\rangle}{\partial t} = \sum_{K'Q'} R_{KQK'Q'}\langle T(J, t)^\dagger_{K'Q'}\rangle \tag{7.5.5}$$

where

$$R_{KQK'Q'} = \sum_{\substack{M'M \\ m'm}} R_{M'Mm'm}(-1)^{2J-M'-m'}[(2K+1)(2K'+1)]^{1/2}$$

$$\times \begin{pmatrix} J & J & K \\ M' & -M & -Q \end{pmatrix}\begin{pmatrix} J & J & K' \\ m' & -m & -Q' \end{pmatrix} \tag{7.5.6}$$

From the Hermiticity condition (7.5.3) it follows that

$$R_{KQK'Q'} = (-1)^{Q+Q'}R_{K-QK'-Q'} \tag{7.5.7}$$

Equation (7.5.5) describes the time evolution of state multipoles in the presence of interactions which causes relaxation. An example of this which is of particular interest is that in which the atomic system under consideration is in an environment which is, on the average, *isotropic*. Isotropic conditions often prevail when a polarized ensemble relaxes toward a random ensemble. It was shown in Chapter 4 that a rotationally invariant interaction cannot alter the rank K and component Q of the tensors. Because the relaxation processes are independent of the choice of the quantization axis the relaxation rates must then be independent on Q. This gives the symmetry condition:

$$R_{KQK'Q'} = -\gamma_K \delta_{K'K}\delta_{Q'Q} \tag{7.5.8}$$

where γ_K is the relaxation rate for all components of the tensor of rank K. Equation (7.5.7) implies that γ_K is *real*,

$$\gamma_K^* = \gamma_K \tag{7.5.9a}$$

and it can be shown that

$$\gamma_K \geq 0 \tag{7.5.9b}$$

Using the symmetry condition (7.5.8) Eq. (7.5.5) can be simplied and all $(2K+1)$ components of a tensor of rank K decay at the same rate:

$$\frac{\partial \langle T(J,t)_{KQ}^\dagger \rangle}{\partial t} = -\gamma_K \langle T(J,t)_{KQ}^\dagger \rangle \tag{7.5.10a}$$

or

$$\langle T(J,t)_{KQ}^\dagger \rangle = \langle T(J,0)_{KQ}^\dagger \rangle \exp(-\gamma_K t) \tag{7.5.10b}$$

Since the monopole $\langle T(J,t)_{00} \rangle$ is proportional to the trace of the density matrix (which is constant if no atoms are removed from the multiplet J) it follows that

$$\gamma_0 = 0 \tag{7.5.11}$$

and hence all multipoles with $K > 0$ will vanish in the course of time, and at thermal equilibrium all substates will be equally populated. If the multiplet under consideration is not the ground state then radiative decay must also be taken into account.

In conclusion we have seen that *under isotropic conditions, each multipole is decoupled from all other multipoles and relaxes with a characteristic relaxation rate* γ_K. The number of independent rates is therefore reduced to $(2J + 1)$. This number is still large if J is high but often not all of the parameters are of interest. A considerable simplification occurs, for example, when the atoms have been excited by dipole radiation. In this case only atomic orientation and alignment can be produced (see Chapter 5) and, irrespective of the value of J, it is only necessary to consider the corresponding relaxation rates γ_1 and γ_2. This is the case in most optical pumping experiments.

In some cases the relaxation process is not isotropic but axially symmetric with respect to a preferred axis, for example, if the relaxation process is itself anisotropic or an external field is present as in the case of magnetic resonance experiments. When the high-frequency field is switched off the atoms then relax in the presence of the static magnetic field which is producing the energy difference between the substates.

It has been shown in Chapter 4 that the component Q of the state multipoles is conserved in an interaction with axial symmetry. In this case Eq. (7.5.5) becomes

$$\frac{\partial \langle T(J, t)^{\dagger}_{KQ}\rangle}{\partial t} = R_{KQK'Q}\langle T(J, t)^{\dagger}_{K'Q}\rangle \tag{7.5.12}$$

This expression shows that tensors with different rank and the same component are mixed by the relaxation processes. In particular, orientation and alignment parameters with the same Q will combine with each other.

The state multipole formalism is of considerable interest for the description of relaxation in atomic and nuclear physics. For a more detailed discussion and applications to particular cases we refer, for example, to the reviews by Omont (1977), Baylis (1979), and to the references cited in these papers.

7.6. The Liouville Formalism

The purpose of this section is to describe some mathematical techniques that are particularly useful in nonequilibrium quantum statistics. These techniques are connected with the Liouville representation of density matrices.

The elements of the Hilbert space are the state vectors $|\psi\rangle$. Consider now the set of all linear operators $A, B, \ldots$ acting on the states $|\psi\rangle$. Any linear combination of linear operators is also a linear operator. Thus the set of all linear operators span another linear space, called the *Liouville space*, if an inner product is defined by the relation

▶
$$(A|B) = \operatorname{tr} A^{\dagger} B \qquad (7.6.1)$$

We will use the notation $|A), |B), \ldots$ to emphasize that these operators are to be considered elements in the Liouville space.

Consider a set of basis vectors $|m'\rangle, |m\rangle, \ldots$ in Hilbert space. A basis in the Liouville space is then represented by the set of all operators $|m'\rangle\langle m|$ which can be obtained by combining all elements of the set $|m\rangle$. Following Gabriel (1969) and using the "Dirac" notation $|m'm)$ for $|m'\rangle\langle m|$ and $(m'm|$ for $|m\rangle\langle m'|$ and using Eq. (7.6.1) we find the orthogonality relation

$$(m'm|n'n) = \operatorname{tr}\{|m\rangle\langle m'|n'\rangle\langle n|\} = \delta_{m'n'}\delta_{mn} \qquad (7.6.2)$$

and the completeness relation

$$\sum_{m'm} |m'm)(m'm| = \mathbf{1} \qquad (7.6.3)$$

where $\mathbf{1}$ is the unit operator in Liouville space. It follows then that

$$(m'm|A) = \operatorname{tr}[|m\rangle\langle m'|A] = \langle m'|A|m\rangle \qquad (7.6.4)$$

that is, the inner product of any Liouville vector A with the basis vectors $|m'm)$ is given by the usual matrix element $\langle m'|A|m\rangle = A_{m'm}$ in Hilbert space.

When the angular symmetries of the system under consideration are important it is convenient to use the elements $|T(J'J)_{KQ})$ as basis vectors. This set is orthonormal,

$$(T(J'J)_{K'Q'}|T(J'J)_{KQ}) = \operatorname{tr}\{T(J'J)_{K'Q'}^{\dagger} T(J'J)_{KQ}\}$$

$$= \delta_{J'J}\delta_{K'K}\delta_{Q'Q} \qquad (7.6.5)$$

in accordance with Eq. (4.2.24) and complete

$$\sum_{J'JKQ} |T(J'J)_{KQ})(T(J'J)_{KQ}| = \mathbf{1} \qquad (7.6.6)$$

Applying the operator (7.6.6) to the density matrix, considered as a vector $|\rho)$ in the Liouville space, gives immediately the expansion:

$$|\rho) = \sum_{J'JKQ} |T(J'J)_{KQ})(T(J'J)_{KQ}|\rho) \qquad (7.6.7)$$

The state multipoles are then interpreted as the inner products

$$\langle T(J'J)^{\dagger}_{KQ} \rangle = (T(J'J)_{KQ} | \rho) \qquad (7.6.8)$$

which can be transformed into the usual form (4.3.5) by using Eq. (7.6.1). Equations (7.6.7) and (7.6.8) correspond to the expansion (4.3.4).

To make calculations in the Liouville space it is necessary to introduce operators $\hat{Q}$ which transform a vector $|A)$ into another vector

$$\hat{Q}|A) = |\hat{Q}A) \qquad (7.6.9)$$

The operators $\hat{Q}$ are often referred to as "superoperators." In an arbitrary basis

$$(m'm|\hat{Q}|A) = \sum_{n'n} (m'm|\hat{Q}|n'n)(n'n | A)$$

$$= \sum_{n'n} Q_{m'mn'n}A_{n'n} \qquad (7.6.10)$$

where the completeness relation (7.6.3) and Eq. (7.6.4) has been used. The elements of any superoperator are thus characterized by four indices.

As an example consider the general relaxation equation (7.5.1):

$$\dot{\rho}_{m'm} = \sum_{n'n} R_{m'mn'n}\rho_{n'n}$$

Interpretating the matrix elements as inner products between Liouville vectors according to Eq. (7.6.4) we write

$$(m'm | \dot{\rho}) = \sum_{n'n} (m'm|\hat{R}|n'n)(n'n | \rho)$$

Substitution of the identity operator (7.6.3) yields then

$$(m'm | \dot{\rho}) = (m'm|\hat{R}|\rho)$$

or

$$|\dot{\rho}) = \hat{R}|\rho) \qquad (7.6.11)$$

where $\hat{R}$ is the relaxation superoperator.

Of particular importance in nonequilibrium quantum statistics is the *Liouville operator* $\hat{L}$, defined for a given Hamiltonian H by

▶ $$\hat{L}|A) = (1/\hbar)|[H, A]) \qquad (7.6.12)$$

for any operator A where $[H, A]$ denotes the usual commutator in Hilbert space. A convenient basis $|i, j)$ for $\hat{L}$ can be constructed from the eigenstates

$|i\rangle, |j\rangle, \ldots$ of the Hamiltonian. In this basis

$$\hat{L}|ij) = \hat{L}|i\rangle\langle j|$$
$$= (1/\hbar)[H, |i\rangle\langle j|])$$
$$= \omega_{ij}|ij) \tag{7.6.13}$$

which shows that the eigenvalues of the Liouville operator are identical to the possible frequencies ω_{ij} of the system.

The equation of motion (2.4.16) of the density matrix can be rewritten in the Liouville notation:

$$|\dot{\rho}) = -(i/\hbar)[H, \rho])$$

or

▶ $$|\dot{\rho}) = -i\hat{L}|\rho) \tag{7.6.14}$$

The formal solution of Eq. (7.6.14) is

$$|\rho(t)) = |\rho(0)) \exp(-i\hat{L}t) \tag{7.6.15}$$

with the time evolution superoperator

$$\hat{U}(t) = \exp(-i\hat{L}t) \tag{7.6.16}$$

Equation (7.6.15) should be compared with the traditional form (2.4.16).

Equations (7.6.7) and (7.6.15) allow a compact representation of the perturbation coefficients defined by Eq. (4.7.6). Assuming that the system of interest is described by a density matrix

$$|\rho(0)) = \sum_{KQ} \langle T(J)_{KQ}^{\dagger}\rangle |T(J)_{KQ})$$

at time $t = 0$ and by

$$|\rho(t)) = \sum_{kq} \rangle T(J, t)_{kq}^{\dagger}\rangle |T(J)_{kq})$$

at time t we obtain from Eqs. (7.6.8) and (7.6.15)

$$\langle T(J, t)_{kq}^{\dagger}\rangle = (T(J)_{kq}|\rho(t))$$
$$= (T(J)_{kq}|\hat{U}(t)|\rho(0))$$
$$= \sum_{KQ} \langle T(J)_{KQ}^{\dagger}\rangle G(t)_{Kk}^{Qq} \tag{7.6.17}$$

where the perturbation coefficient in Liouville notation is represented by

$$G(t)_{Kk}^{Qq} = (T(J)_{kq}|\hat{U}(t)|T(J)_{KQ}) \tag{7.6.18}$$

The formalism presented here has been applied by various authors to the theory of angular correlations perturbed by relaxation effects [see, for example, the papers by Gabriel (1969) and Bosse and Gabriel (1974)].

The real advantage of the Liouville operator appears in the resolvent form. The resolvent method, which adopts the concepts and techniques of scattering theory to the Liouville representation of density matrices, allows to represent the formalism in a compact form. For an introduction into this method we refer to the original paper by Zwanzig (1960).

7.7. Linear Response of a Quantum System to an External Perturbation

Physical problems are concerned with the determination of the unknown properties of a system. In order to do this one lets an external agent act on the system and observes the reaction of the system. That is, the observer puts a question to the system and the system responses. Starting with such general consideration the response formalism has been developed. This formalism was first applied by Kubo to the theory of irreversible processes in order to study transport phenomena, for example, the effect of external forces on equilibrium systems disturbing the equilibrium state and causing the system to conduct heat or electricity, or otherwise respond to the stimulus. In the present section a brief introduction into this formalism will be given.

Let a quantum system described by a density matrix $\rho(t)$ be subjected to the action of an external perturbation $V(t)$. In the theory of irreversible processes it is usually assumed that the system was in statistical equilibrium with a heat bath in the remote past $(t \to -\infty)$, which is expressed by the initial condition

$$\rho(t) \to \rho_0 \qquad (7.7.1)$$

for $t \to -\infty$, where ρ_0 is the equilibrium density matrix of the system. The time evolution of the density matrix is given by the Liouville equation (2.4.16). Assuming that $V(t)$ is sufficiently small then in first-order perturbation theory the solution of the Liouville equation can be written in the form

$$\rho(t) = \rho_0 - (i/\hbar) \int_{-\infty}^{t} dt' \exp\left[iH_0(t' - t)/\hbar\right][V(t'), \rho_0] \exp\left[-iH_0(t' - t)/\hbar\right]$$
$$(7.7.2)$$

which is obtained by transforming Eq. (2.4.43) back into the Schrödinger picture by using Eqs. (2.4.25) and (2.4.37). In this approximation the change in the expectation value $\langle A \rangle$ of an operator A is given by

$$\Delta\langle A(t)\rangle = \text{tr}\,\{\rho(t)A\} - \text{tr}\,\rho_0 A$$

$$= -\frac{i}{\hbar}\,\text{tr}\,\left\{\int_{-\infty}^{t} \exp\left[\frac{iH_0(t'-t)}{\hbar}\right][V(t'),\rho_0]\exp\left[-\frac{iH_0(t'-t)}{\hbar}\right]A\,dt'\right\}$$

$$(7.7.3)$$

The quantity $\Delta\langle A(t)\rangle$ can be regarded as the first-order response of the system to an external perturbation. It should be noted that the Hamiltonian $H_0 + V(t)$ refers only to the system itself and the influence of the heat bath is not taken into account.

Suppose that $V(t)$ can be represented in the form

$$V(t) = -f(t)B \qquad (7.7.4)$$

where $f(t)$ is an external driving force (for example, an electric field) and B an operator on the system (for example, the dipole operator). Inserting Eq. (7.7.4) into Eq. (7.7.3) gives

$$\Delta\langle A(t)\rangle = (i/\hbar)\int_{-\infty}^{t} dt'\,\text{tr}\,\{[B,\rho_0]A(t'-t)\}f(t')$$

$$- (i/\hbar)\int_{-\infty}^{t}\text{tr}\,\{\rho_0[A(t'-t),B]\}f(t')\,dt' \qquad (7.7.5)$$

where the cyclic property of the trace has been used and where

$$A(t'-t) = \exp\{[-iH_0(t'-t)]/\hbar\}A\,\exp\,[iH_0(t'-t)/\hbar] \qquad (7.7.6)$$

The upper limit of integration in Eq. (7.7.5) can be extended to infinity by introducing the *Green's function*

▶ $$\langle\!\langle A(t)B(t')\rangle\!\rangle = -(i/\hbar)\theta(t-t')\,\text{tr}\,\{\rho_0[A(t),B(t')]\} \qquad (7.7.7)$$

where $\theta(t-t')$ is the step function. $\Delta\langle A(t)\rangle$ can then be written in the form

▶ $$\Delta\langle A(t)\rangle = \int_{-\infty}^{\infty}\langle\!\langle A(t'-t)B\rangle\!\rangle f(t')\,dt' \qquad (7.7.8)$$

Equation (7.7.8) shows that the effect of an external perturbation on the mean values of observables can be described by Green's functions coupling the observed quantity with the perturbation.

An interpretation of the Green's function can be obtained by considering a unit pulse at time t_1, that is, by letting $f(t')$ become $\delta(t'-t_1)$ [where $\delta(t'-t_1)$ denotes Dirac's delta function]. In this case it follows from Eq. (7.7.8) that

$$\Delta\langle A(t)\rangle = \langle\!\langle A(t_1-t)B\rangle\!\rangle \qquad (7.7.9)$$

The Green's function $\langle\langle A(t' - t)B\rangle\rangle$ is therefore the change $\Delta\langle A(t)\rangle$ by the time t due to a unit pulse at time t'. Equation (7.7.8) can then be interpreted as the linear superposition of responses produced by pulses at times t' with amplitudes $f(t')$. The range of values of t' in Eq. (7.7.8) is $t' < t$ (otherwise the step function vanishes). The response has therefore a *causal* character since only the effects of the perturbation at past moments of the time are taken into account. For this reason the quantity (7.7.7) is called the *retarded Green's function*.

Equation (7.7.8) is called the *Kubo formula* for the linear response of a system. The important point to note is that this equation expresses nonequilibrium properties in terms of averages over equilibrium states. One could also define the nonlinear response of a system to an external agent. In this case, however, the Green's functions are no longer properties of the unperturbed system.

As a special case consider a periodic perturbation

$$V(t) = -V_0 \exp(-i\omega t + \varepsilon t)B \qquad (7.7.10)$$

where V_0 is the amplitude and ε an infinitesimally small quantity which ensures $V(t) \to 0$ for $t \to -\infty$. In the case of the periodic perturbation (7.7.10) the equation (7.7.8) takes the form

$$\Delta\langle A(t)\rangle = V_0 \exp(-i\omega t + \varepsilon t) \int_{-\infty}^{\infty} dt' \,\langle\langle A(t' - t)B\rangle\rangle$$

$$\times \exp[-i\omega(t' - t) + \varepsilon(t' - t)]$$

$$= V_0 \exp(-i\omega t + \varepsilon t) \int_{-\infty}^{\infty} \langle\langle A(\tau)B\rangle\rangle \exp(-i\omega\tau + \varepsilon\tau)\,d\tau \qquad (7.7.11a)$$

$$= V_0 \exp(-i\omega t + \varepsilon t)\langle\langle AB\rangle\rangle_\omega \qquad (7.7.11)$$

where $\tau = t' - t$ and $\langle\langle AB\rangle\rangle_\omega$ is defined by the integral in Eq. (7.7.11a). The *generalized susceptibility* $\chi(\omega)$ which describes the influence of the periodic perturbation (7.7.10), is defined by

$$\Delta\langle A(t)\rangle = \chi(\omega)V_0 \exp(-i\omega t + \varepsilon t) \qquad (7.7.12)$$

Comparing Eqs. (7.7.11) and (7.7.12) we obtain

$$\chi(\omega) = \langle\langle AB\rangle\rangle_\omega \qquad (7.7.13)$$

This is Kubo's formula for the generalized susceptibility.

The equations derived here can be used as a starting point for the investigation of transport phenomena. Under appropriate conditions it is possible to connect the response formalism with the Onsager theory of irreversible processes. Assuming that the external perturbation is so weak that the discussion can be restricted to first-order perturbation theory it has been shown that transport coefficients can be evaluated using an equilibrium density matrix. For example, the electric conductivity is directly related to the response of a system to an external field and this response in turn is related to time correlation functions. A discussion of these topics falls outside the scope of this book. For a detailed account with many applications the reader is referred particularly to the treatment given by Zubarev (1974).

Appendixes

Appendix A: The Direct Product

An important quantity in matrix algebra is the direct product $C = A \times B$ of two matrices A and B, where each element of C is formed by replacing each element a_{ij} of A by the matrix $a_{ij}B$. Thus if A is N dimensional and B n dimensional then C is an $N \times n$-dimensional matrix. For example, if

$$A = \begin{pmatrix} a_{12} & a_{12} \\ a_{21} & a_{22} \end{pmatrix}, \qquad B = \begin{pmatrix} b_{11} & b_{12} \\ b_{21} & b_{22} \end{pmatrix}$$

then the direct product $A \times B$ is given by the four-dimensional matrix

$$A \times B = \begin{pmatrix} a_{11}B & a_{12}B \\ a_{21}B & a_{22}B \end{pmatrix} \tag{A1a}$$

where each "element" $a_{ij}B$ stands for the two-dimensional matrix

$$a_{ij}B = \begin{pmatrix} a_{ij}b_{11} & a_{ij}b_{12} \\ a_{ij}b_{21} & a_{ij}b_{22} \end{pmatrix} \tag{A1b}$$

It can be shown that, if A and C are $m \times m$ matrices and B and D $n \times n$ matrices, then the usual matrix product of $A \times B$ and $C \times D$ is given by

$$(A \times B) \cdot (C \times D) = (AC) \times (BD) \tag{A2a}$$

An important trace relation is

$$\text{tr} (A \times B) = \text{tr} A \, \text{tr} B \tag{A2b}$$

Application of the relations (A2) enables the use of explicit matrix representations to be avoided in most calculations.

The definition (A1) can also be applied to row vectors which can be considered as matrices with one row only. For example, when the spin states are written in the standard representation, the direct product of the spin-1 state $|+1\rangle$ and the spin-1/2 state $|-1/2\rangle$ is given by

$$\begin{pmatrix} 1 \\ 0 \\ 0 \end{pmatrix} \times \begin{pmatrix} 0 \\ 1 \end{pmatrix} = \begin{pmatrix} 0 \\ 1 \\ 0 \\ 0 \\ 0 \\ 0 \end{pmatrix} \tag{A3}$$

which we will write in the form: $|-1\rangle|-1/2\rangle = |1, -1/2\rangle$. (A4)

More generally, consider two linear spaces R and r, spanned by basis vectors $|N\rangle$ and $|n\rangle$, respectively [that is, any state vector in the space $R(r)$ can be written as a linear combination of the states $|N\rangle$ ($|n\rangle$)]. The combined space can be spanned by the set of all direct product states

$$|N, n\rangle = |N\rangle|n\rangle \tag{A5}$$

that is, by all, possible pairs formed from all basis vectors $|N\rangle$ and $|n\rangle$.

For example, an ensemble of spin-1 particles may be in the state $|+1\rangle$ and an ensemble of spin-1/2 particles in the state $|-1/2\rangle$. When both systems are well separated and not interacting the state of the combined system is represented by the direct product $|+1, -1/2\rangle$. This simple representation does not apply when the two systems interact (see Section 3.1). However, any state vector $|\psi\rangle$ representing the state of the coupled system, can always be written as a sum of direct products

$$|\psi\rangle = \sum_{Mm} a(M, m)|M\rangle|m\rangle \tag{A6}$$

with $M = \pm1, 0$ and $m = \pm1/2$.

The direct product states have the following important properties. A scalar product is defined by

$$\langle N', n'|N, n\rangle = \langle N'|N\rangle\langle n'|n\rangle$$

The matrix elements of an operator $Q(R)$, acting only on the space R, are given by

$$\langle N', n'|Q(R)|N, n\rangle = \langle N'|Q(R)|N\rangle\langle n'|n\rangle \tag{A7}$$

and for a direct product

$$\langle N', n'|Q(R) \times Q(r)|N, n\rangle = \langle N'|Q(R)|N\rangle\langle n'|Q(r)|n\rangle \tag{A8}$$

Consider now a mixture of states $|N, n\rangle$ represented by a density matrix

$$\rho = \sum_{Nn} W_{Nn} |N, n\rangle\langle N, n| \qquad (A9)$$

where W_{Nn} is the probability of finding the system in the state $|N, n\rangle = |N\rangle|n\rangle$. The two systems are *uncorrelated* if

$$W_{Nn} = W_N W_n \qquad (A10)$$

that is, when the probability of finding one system in a state $|N\rangle$ is independent of the probability of finding the other system in a state $|n\rangle$. When Eq. (A10) holds then, from Eqs. (A5) and (A9),

$$\rho = \left(\sum_N W_N |N\rangle\langle N| \right)\left(\sum_n W_n |n\rangle\langle n| \right)$$

$$= \rho(N) \times \rho(n) \qquad (A11)$$

That is, *in the special case of uncorrelated systems the total density matrix is represented by the direct product of the individual matrices.*

As an example, consider two ensembles of particles with spins S_1 and S_2, respectively. Before any interaction the two systems are uncorrelated and represented by density matrices $\rho(S_1)$ and $\rho(S_2)$, respectively. The combined system then is characterized by the density matrix

$$\rho_{ih} = \rho(S_1) \times \rho(S_2)$$

Expanding the density matrices $\rho(S_1)$ and $\rho(S_2)$ in terms of spin tensors as discussed in Section 4.4 we obtain

$$\rho_{in} = \left[\sum_{KQ} \langle T(S_1)^{\dagger}_{KQ}\rangle T(S_1)_{KQ} \right]\left[\sum_{kq} \langle T(S_2)^{\dagger}_{kq}\rangle T(S_2)_{kq} \right]$$

$$= \sum_{\substack{KQ \\ kq}} \langle T(S_1)^{\dagger}_{KQ}\rangle\langle T(S_2)^{\dagger}_{kq}\rangle [T(S_1)_{KQ} \times T(S_2)_{kq}] \qquad (A12)$$

Using Eqs. (A2b), (4.2.24), and (4.2.25) we obtain for the trace

$$\operatorname{tr}[\rho_{in} \cdot T(S_1)_{K'Q'} \times \mathbf{1}] = \sum_{\substack{KQ \\ kq}} \langle T(S_1)^{\dagger}_{KQ}\rangle\langle T(S_2)^{\dagger}_{kq}\rangle$$

$$\cdot \operatorname{tr}[T(S_1)_{KQ} \times T(S_2)_{kq}][T(S_1)^{\dagger}_{K'Q'} \times \mathbf{1}]$$

$$= [1/(2S_2 + 1)^{1/2}]\langle T(S_1)^{\dagger}_{K'Q'}\rangle \cdot \langle T(S_2)_{00}\rangle$$

Application of Eq. (4.3.14) then finally gives

$$T(S_1)^{\dagger}_{K'Q'}\rangle = \operatorname{tr}\rho_{in}[T(S_1)^{\dagger}_{K'Q} \times \mathbf{1}] \qquad (A13)$$

Similarly, the spin tensors characterizing the second system only are given by

$$\langle T(S_2)^\dagger_{kq} \rangle = \text{tr } \rho_{\text{in}} [\mathbf{1} \times T(S_2)^\dagger_{kq}] \tag{A14}$$

Equations (A12), (A13), and (A14) are used, for example, in scattering theory to describe the initial state of polarized particles.

Appendix B: State Multipoles for Coupled Systems

Consider two interacting systems with angular momenta J and I, respectively. The two systems may consist of, for example, two different ensembles of particles such as electrons with spin $I = 1/2$ and atoms with spin J or of two different characteristics of the same state (for example, an atomic state with electronic angular momentum J and nuclear spin I).

State multipoles describing the coupled system can be constructed by first coupling the states $|JM\rangle$ and $|Im\rangle$ to eigenstates of the total angular momentum operator F and then using these eigenstates to construct tensor operators $T(F'F)_{KQ}$ using Eq. (4.2.3) with state multipoles corresponding to Eq. (4.3.3).

It is often more convenient to represent the total density matrix ρ in a different way. We take the set of all tensor operators $T(J)_{KQ}$ and $T(I)_{kq}$ which describe the separate systems (with $K \leq 2J$ and $k \leq 2I$) and construct the set of all direct products $T(J)_{KQ} \times T(I)_{kq}$ as in Appendix A. Any operator acting on the composite space, spanned by the direct products $|JM\rangle|Im\rangle$, can then be expanded in terms of this set. Hence

$$\rho = \sum_{\substack{KQ \\ kq}} \langle T(J)^\dagger_{KQ} \times T(I)^\dagger_{kq} \rangle [T(J)_{KQ} \times T(I)_{kq}] \tag{B1}$$

The state multipoles are obtained from Eq. (B1) calculating the trace

$$\langle T(J)^\dagger_{KQ} \times T(I)^\dagger_{kq} \rangle = \text{tr } \rho [T(J)^\dagger_{KQ} \times T(I)^\dagger_{kq}] \tag{B2}$$

according to Appendix A.

When the two systems are uncorrelated

$$\langle T(J)^\dagger_{KQ} \times T(I)^\dagger_{kq} \rangle = \langle T(J)^\dagger_{KQ} \rangle \langle T(I)^\dagger_{kq} \rangle \tag{B3}$$

as follows from Appendix A.

In many cases the parameters

$$\langle T(J)^\dagger_{KQ} \times \mathbf{1} \rangle = (2I + 1)^{1/2} \langle T(J)^\dagger_{KQ} \times T(I)_{00} \rangle \tag{B4}$$

are of particular interest. In Eq. (B4) Eq. (4.2.14) has been used, and $\mathbf{1}$ is the $(2I + 1)$-dimensional unit matrix. Using Eq. (A8b) it can readily be shown that

$$\langle T(J)_{KQ}^{\dagger} \times 1 \rangle = \text{tr } \rho \cdot [T(J)_{KQ} \times \mathbf{1}]$$

$$= \text{tr } [\rho(J)T(J)_{KQ}^{\dagger}] \tag{B5}$$

where $\rho(J)$ is the reduced density matrix describing the J system alone

$$\langle JM'|\rho(J)|JM \rangle = \sum_m \langle JM', Im|\rho|JM, Im \rangle$$

Hence, when only the J system is of interest and the I system undetected, only the set of multipoles $\langle T(J_{KQ}^{\dagger} \times 1 \rangle = \langle T(J)_{KQ}^{\dagger} \rangle$ need be considered. Similarly, when only the I system is observed then the parameters of interest are the multipoles $\langle 1 \times T(I)_{kq}^{\dagger} \rangle = \langle T(I)_{kq}^{\dagger} \rangle$. Examples of this are given in Section 4.7. As another example, consider scattering experiments with polarized particles of spin J and I, respectively. The set of all spin tensors $\langle T(J)_{KQ}^{\dagger} \times 1 \rangle$ and $\langle 1 \times T(I)_{kq}^{\dagger} \rangle$ can be used to characterize the polarization states of the J and I systems, respectively, when the other one is either undetected or unpolarized. When polarization measurements are performed on both systems in coincidence then some or all of the parameters with both K and k nonzero must also be considered. Using Eq. (E5) and expressing ρ_{out} according to Eq. (B2) and ρ_{in} according to Eq. (A12) the spin tensors of the final particles can then be related to those of the initial ones.

Finally, we give the relation describing the transformation between the "coupled" tensors $T(F'F)_{K'Q'}$ and the "uncoupled" operators $T(J)_{KQ} \times T(I)_{kq}$:

$$T(F'F)_{K'Q'} = \sum_{\substack{KQ \\ kq}} [(2K + 1)(2k + 1)(2F' + 1)(2F + 1)]^{1/2}(KQ, kq|K'Q')$$

$$\times \begin{Bmatrix} K & k & K' \\ J & I & F' \\ J & I & F \end{Bmatrix} T(J)_{KQ} \times T(I)_{kq} \tag{B6}$$

where $\{\cdots\}$ denotes a $9j$ symbol. The inverse relation can be obtained by using the orthogonality properties of the $9j$ symbol:

$$T(J)_{KQ} \times T(I)_{kq}$$

$$= \sum_{\substack{FF' \\ K'Q'}} [(2K + 1)(2k + 1)(2F' + 1)(2F + 1)]^{1/2}(KQ, kq|K'Q')$$

$$\times \begin{Bmatrix} K & k & K' \\ J & I & F' \\ J & I & F \end{Bmatrix} T(F'F)_{K'Q'} \tag{B7}$$

As a special case, for $k = 0$ Eq. (B7) gives

$$T(J)_{KQ} \times \mathbf{1} = (2I + 1)^{1/2}[T(J)_{KQ} \times T(I)_{00}]$$

$$= \sum_{F'F} [(2F' + 1)(2F + 1)]^{1/2}(-1)^{F+J+K+I}\begin{Bmatrix} F' & F & K \\ J & J & I \end{Bmatrix}$$

(B8)

A similar relation holds for the tensor operators $1 \times T(I)_{kq}$ describing only the I system with the J system undetected.

Appendix C: Formulas from Angular Momentum Theory

Clebsch–Gordan Coefficients

$$\sum_{JM} (J_1M'_1, J_2M'_2|JM)(J_1M_1, J_2M_2|JM) = \delta_{M'_1M_1}\delta_{M'_2M_2} \quad \text{(C1a)}$$

$$\sum_{M_1M_2} (J_1M_1, J_2M_2|J'M')(J_1M_1, J_2M_2|JM) = \delta_{J'J}\delta_{MM'} \quad \text{(C1b)}$$

Symmetry properties:

$(J_1M_1, J_2M_2|JM)$

$$= (-1)^{J_1+J_2-J}(J_1 - M_1, J_2 - M_2|J - M) \quad \text{(C2a)}$$

$$= (-1)^{J_1+J_2-J}(J_2M_2, J_1M_1|JM) \quad \text{(C2b)}$$

$$= [(2J + 1)/(2J_2 + 1)]^{1/2}(-1)^{J_1-M_1}(J_1M_1, J - M|J_2 - M_2) \quad \text{(C2c)}$$

$$= [(2J + 1)/(2J_1 + 1)]^{1/2}(-1)^{J_2+M_2}(J - M, J_2M_2|J_1 - M_1) \quad \text{(C2d)}$$

3j Symbols

Definition:

$$\begin{pmatrix} J_1 & J_2 & J \\ M_1 & M_2 & M \end{pmatrix} = [1/(2J + 1)^{1/2}](-1)^{J_1-J_2-M}(J_1M_1, J_2M_2|J - M) \quad \text{(C3)}$$

Orthogonality relations:

$$\sum_{JM} (2J + 1)\begin{pmatrix} J_1 & J_2 & J \\ M_1 & M_2 & M \end{pmatrix}\begin{pmatrix} J_1 & J_2 & J \\ M'_1 & M'_2 & M \end{pmatrix} = \delta_{M'_1M_1}\delta_{M_2M'_2} \quad \text{(C4a)}$$

$$\sum_{M_1M_2} \begin{pmatrix} J_1 & J_2 & J' \\ M_1 & M_2 & M' \end{pmatrix}\begin{pmatrix} J_1 & J_2 & J \\ M_1 & M_2 & M \end{pmatrix} = [1/(2J + 1)]\delta_{JJ'}\delta_{MM'} \quad \text{(C4b)}$$

Symmetry properties: The $3j$ symbol is invariant under cyclic permutations of its columns and multiplied by $(-1)^{J_1+J_2+J}$ by noncyclic ones. In particular,

$$\begin{pmatrix} J_1 & J_2 & J \\ M_1 & M_2 & M \end{pmatrix} = \begin{pmatrix} J_2 & J & J_1 \\ M_2 & M & M_1 \end{pmatrix} = \begin{pmatrix} J & J_1 & J_2 \\ M & M_1 & M_2 \end{pmatrix} \tag{C5a}$$

$$\begin{pmatrix} J_1 & J_2 & J \\ M_1 & M_2 & M \end{pmatrix} = (-1)^{J_1+J_2+J}\begin{pmatrix} J_2 & J_1 & J \\ M_2 & M_1 & M \end{pmatrix} \tag{C5b}$$

$$= (-1)^{J_1+J_2+J}\begin{pmatrix} J_1 & J_2 & J \\ -M_1 & -M_2 & -M \end{pmatrix} \tag{C5c}$$

Special case:

$$\begin{pmatrix} J_1 & J_2 & 0 \\ M_1 & -M_2 & 0 \end{pmatrix} = \frac{(-1)^{J_1-M_1}}{(2J_1+1)^{1/2}}\delta_{J_1 J_2}\delta_{M_1 M_2} \tag{C6}$$

6j Symbols

Definition:

$$\begin{Bmatrix} J_1 & J_2 & J_3 \\ j_1 & j_2 & j_3 \end{Bmatrix} =$$

$$\sum_{\text{all } M_i, m_i} (-1)^{\Sigma J_i + \Sigma j_i + \Sigma m_i}\begin{pmatrix} J_1 & J_2 & J_3 \\ -M_1 & -M_2 & -M_3 \end{pmatrix}\begin{pmatrix} J_1 & j_2 & j_3 \\ M_1 & m_2 & -m_3 \end{pmatrix}$$

$$\times \begin{pmatrix} j_1 & J_2 & j_3 \\ -m_1 & M_2 & m_3 \end{pmatrix}\begin{pmatrix} j_1 & j_2 & J_3 \\ m_1 & -m_2 & M_3 \end{pmatrix} \tag{C7}$$

Symmetries: The $6j$ symbol is invariant for interchange of any two columns, and for interchange of the upper and lower arguments in each of any two columns, for example

$$\begin{Bmatrix} J_1 & J_2 & J_3 \\ j_1 & j_2 & j_3 \end{Bmatrix} = \begin{Bmatrix} J_1 & J_3 & J_2 \\ j_1 & j_3 & j_2 \end{Bmatrix} = \begin{Bmatrix} j_1 & j_2 & J_3 \\ J_1 & J_2 & j_3 \end{Bmatrix} \tag{C8}$$

Contraction:

$$\sum_{M_1 M_2 M_3} (-1)^{J_1+J_2+J_3+M_1+M_2+M_3}\begin{pmatrix} J_1 & J_2 & j_3 \\ M_1 & -M_2 & m_3 \end{pmatrix}\begin{pmatrix} J_2 & J_3 & j_1 \\ M_2 & -M_3 & m_1 \end{pmatrix}$$

$$\times \begin{pmatrix} J_3 & J_1 & j_2 \\ M_3 & -M_1 & m_2 \end{pmatrix} = \begin{pmatrix} j_1 & j_2 & j_3 \\ m_1 & m_2 & m_3 \end{pmatrix}\begin{Bmatrix} j_1 & j_2 & j_3 \\ J_1 & J_2 & J_3 \end{Bmatrix} \tag{C9}$$

Orthogonality:

$$\sum_{J} (2J + 1)(2J'' + 1)\begin{Bmatrix} J_1 & J_2 & J' \\ J_3 & J_4 & J \end{Bmatrix}\begin{Bmatrix} J_3 & J_2 & J \\ J_1 & J_4 & J'' \end{Bmatrix} = \delta_{J'J''} \tag{C10}$$

Special value:

$$\begin{Bmatrix} J_1 & J_2 & J_3 \\ 0 & J_3 & J_2 \end{Bmatrix} = \frac{(-1)^{J_1+J_2+J_3}}{[(2J_2 + 1)(2J_3 + 1)]^{1/2}} \tag{C11}$$

Rotation Matrix Elements

Definition:

$$D(\gamma\beta\alpha)^{(J)}_{M'M} = \exp{(iM'\gamma)}d(\beta)^{(J)}_{M'M}\exp{(iM\alpha)} \tag{C12}$$

Symmetries:

$$d(\beta)^{(J)}_{M'M} = d(-\beta)^{(J)}_{MM'}$$
$$= (-1)^{M'-M}d(\beta)^{(J)}_{MM'} = (-1)^{M'-M}d(\beta)^{(J)}_{-M'-M} \tag{C13}$$
$$D(\gamma\beta\alpha)^{(J)*}_{M'M} = (-1)^{M'-M}D(\gamma\beta\alpha)^{(J)}_{-M'-M} \tag{C14}$$

Special values:

$$d(\pi)^{(J)}_{M'M} = (-1)^{J+M}\delta_{M'-M}, \qquad d(0)^{(J)}_{M'M} = \delta_{M'M} \tag{C15}$$

Relation to the spherical harmonics Y_{JM} and Legendre polynomials P_J:

$$D(\gamma\beta\alpha)^{(J)}_{M0} = (-1)^{M}[4\pi/(2J + 1)]^{1/2}Y(\beta\gamma)_{JM} \tag{C16}$$

$$D(\gamma\beta\alpha)^{(J)}_{0M} = [4\pi/(2J + 1)]^{1/2}Y(\beta\alpha)_{JM} \tag{C17}$$

$$D(\gamma\beta\alpha)^{(J)}_{00} = P(\cos\beta)_J \tag{C18}$$

Contraction ($\omega = \gamma\beta\alpha$):

$$D(\omega)^{(J_1)}_{M'_1M_1}D(\omega)^{(J_2)}_{M'_2M_2}$$

$$= \sum_{JMM'} (2J + 1)\begin{pmatrix} J_1 & J_2 & J \\ M'_1 & M'_2 & M' \end{pmatrix}\begin{pmatrix} J_1 & J_2 & J \\ M_1 & M_2 & M \end{pmatrix}D(\omega)^{(J)*}_{M'M} \tag{C17}$$

Orthogonality:

$$\int D(\gamma\beta\alpha)^{(j)*}_{mm'}D(\gamma\beta\alpha)^{(J)}_{MM'}\sin\beta\, d\beta\, d\alpha\, d\gamma = (8\pi^2/(2J + 1))\delta_{Jj}\delta_{mM}\delta_{m'M'}$$

$$\tag{C18}$$

Matrix Elements of Irreducible Tensor Operators

Wigner–Eckart theorem:

$$\langle J'M'|T_{KQ}|JM\rangle = (-1)^{J'-M'}\begin{pmatrix} J' & K & J \\ -M' & Q & M \end{pmatrix}\langle J'\|T_K\|J\rangle \qquad (C19)$$

Reduction for composite systems $(L + S = J, L' + S' = J')$: If the tensor operator T_{KQ} acts only on the system with angular momenta L, L' then

$$\langle (L'S')J'\|T_K\|(LS)J\rangle = (-1)^{L'+S+J+K}[(2J' + 1)(2J + 1)]^{1/2}$$

$$\times \begin{Bmatrix} L' & J' & S \\ J & L & K \end{Bmatrix}\langle L'\|T_K\|L\rangle\,\delta_{SS'} \qquad (C20)$$

Appendix D: The Efficiency of a Measuring Device

The diagonal elements $\langle n|\rho|n\rangle$ of the density matrix are the probabilities that the pure state $|n\rangle$ will be observed in an experimental observation. However, most experimental situations which can be devised will not respond to only one particular pure state and, in general, the detector responds to several states $|n\rangle$ with relative probabilities ("efficiencies") ε_n. The total probability of the response of the apparatus will then be given by

$$W = \sum_n \varepsilon_n \langle n|\rho|n\rangle \qquad (D1)$$

in the representation with basis vectors $|n\rangle$. We introduce the operator

$$\varepsilon = \sum_n \varepsilon_n |n\rangle\langle n| \qquad (D2)$$

which is analogous to the density operator (2.2.1), and Eq. (D1) can then be written in the form

$$W = \text{tr}\,\rho\varepsilon \qquad (D3)$$

ε is called the "efficiency matrix" of the measuring device which completely describes the response of the apparatus. If the apparatus responds only to a single state $|n\rangle$ with certainty (that is, it is a perfect filter), then $\varepsilon = |n\rangle\langle n|$. In this case we can project out of the mixture the definite pure state $|n\rangle$ and Eq. (D1) reduces to

$$W = W_n = \langle n|\rho|n\rangle \qquad (D4)$$

As an example we will consider the measurement of polarization of spin-1/2 particles. The efficiency matrix ε of the polarization filter is a 2×2

matrix which can be expanded in terms of the two-dimensional unit matrix **1** and the Pauli matrices in a similar way as the density matrix in Section 1.1.5:

$$\varepsilon = (1/2)\left(\mathbf{1} + \sum_i Q_i\sigma_i\right) = (1/2)(\mathbf{1} + \mathbf{Q}\boldsymbol{\sigma}) \tag{D5}$$

Transforming to a representation with basis states $|\pm 1/2, z'\rangle$ where ε is diagonal we find

$$\varepsilon = \frac{1}{2}\begin{pmatrix} 1 + Q & 0 \\ 0 & 1 - Q \end{pmatrix} \tag{D6}$$

where $Q = |\mathbf{Q}|$. Hence, in this system, ε can be written in the form

$$\varepsilon = (1/2)(1 + Q)|1/2, z'\rangle\langle 1/2, z'| + (1/2)(1 - Q)|-1/2, z'\rangle\langle -1/2, z'| \tag{D7}$$

which relates the parameters Q to the efficiencies with which the filter responds to the states $|\pm 1/2, z'\rangle$.

Furthermore, it follows from Eqs. (1.1.46) and (D5) that

$$W = \operatorname{tr} \rho\varepsilon = (1/2)(1 + \mathbf{PQ}) \tag{D8}$$

and hence W has its maximum value $W_{\uparrow\uparrow}$ if **P** and **Q** are parallel and its minimum value $W_{\downarrow\downarrow}$ if **P** and **Q** are antiparallel. It follows that the direction of **Q** is that direction in which the polarization filter must be oriented in order to get maximum response.

Thus, in order to determine **P** for a given beam with a polarization filter with a known Q, the orientation of the filter must first be altered until W is a maximum. This direction is then the direction of P. The magnitude $|\mathbf{P}|$ of the polarization vector is then found from a measurement of the values $W_{\uparrow\uparrow}$ and $W_{\downarrow\downarrow}$ at maximum and minimum response and the known value of Q:

$$P = \frac{W_{\uparrow\uparrow} - W_{\downarrow\downarrow}}{Q(W_{\uparrow\uparrow} + W_{\downarrow\downarrow})} \tag{D9}$$

Appendix E: The Scattering and Transition Operators

In scattering theory it is convenient to consider the incoming state of the particles as the state vector $|\psi_{in}\rangle$ at an infinitely remote past when the interaction between the particles can be neglected, and the outgoing state as the state vector $|\psi_{out}\rangle$ at an infinitely late future instant corresponding to such a large distance between the particles for the interaction between them to be neglected again. The S matrix can then be defined by the relation

$$|\psi_{out}\rangle = S|\psi_{in}\rangle \tag{E1}$$

that is, the collision is thought of as a "black box", mathematically represented by S, which transforms the "in" states into the "out" states. When the initial state is represented by the density matrix

$$\rho_{in} = \sum_i W_i |\psi_{in}^{(i)}\rangle\langle\psi_{in}^{(i)}|$$

the density matrix ρ'_{out} describing the final particles, is obtained by operating on ρ_{in} by S and $S^\dagger$:

$$S\rho_{in}S^\dagger = \sum_i W_i S|\psi_{in}^{(i)}\rangle\langle\psi_{in}^{(i)}|S^\dagger$$

$$= \sum_i W_i |\psi_{out}^{(i)}\rangle\langle\psi_{out}^{(i)}|$$

$$\equiv \rho'_{out} \tag{E2}$$

Since one is usually only interested in transitions between different states it is convenient to extract from S the unit operator $\mathbf{1}$ and to define the *transition operator* T by

$$T = S - \mathbf{1} \tag{E3}$$

From Eqs. (E1) and (E2) it follows

$$T|\psi_{in}\rangle = |\psi_{out}\rangle - |\psi_{in}\rangle \tag{E4}$$

All possible transitions (scattering, reactions) in the system will be connected with the dissimilarity between initial and final state, that is, T transforms the "in" state into the scattered state. The interesting part of the density matrix (E2) is then that one which contains the information on the scattered states alone, which is given by

$$\rho_{out} = T\rho_{in}T^\dagger \tag{E5}$$

The central problem of scattering theory is then the determination of T, that is, the determination of all elements of T (see Section 3.5).

References

Abragam, A. (1961): *The Principles of Nuclear Magnetism*, Clarendon Press, Oxford.

Andrä, H. J. (1974): *Physica Scripta* **9**, 252.

Andrä, H. J. (1971), in: *Progress in Atomic Spectroscopy* (eds. Hanle, W., and Kleinpoppen, H.), Plenum Press, New York.

Baylis, W. E. (1979), in: *Progress in Atomic Spectroscopy* (eds. Hanle, W., and Kleinpoppen, H.), Plenum Press, New York.

Bloch, F. (1946): *Phys. Rev.* **70**, 460.

Bloch, F., and Wangsness, R. K. (1952): *Phys. Rev.* **89**, 728.

Blum, K., and Kleinpoppen, H. (1979): *Phys. Rep.* **52**, 203.

Blum, K., and Kleinpoppen, H. (1981): *Adv. At. Mol. Phys.*, to be published.

Born, M., and Wolf, E. (1970): *Principles of Optics*, Pergamon Press, New York.

Bosse, J., and Gabriel, H. (1974): *Z. Phys.* **266**, 283.

Brink, D. M., and Satchler, G. R. (1962): *Angular Momentum*, Clarendon Press, Oxford.

Burke, P. G., and Joachain, C. J. (est. 1982): *Theory of Electron–Atom Scattering*, Plenum Press, New York.

Burns, D., and Hancock, W. H. (1971): *Phys. Rev. Lett.* **27**, 370.

Chow, W., Scully, M. O., and Stoner, W. (1975): *Phys. Rev. A* **11**, 1380.

Cohen-Tannouidji, C. (1962): *Ann. Phys. (Paris)* **7**, 423.

Cohen-Tannouidji, C., and Kastler, A. (1966), in: *Progress in Optics* (ed. Wolf, E.), Vol. 5, North-Holland, Amsterdam.

Corney, A. (1977): *Atomic and Laser Spectroscopy*, Clarendon Press, Oxford.

d'Espagnat, B. (1976): *Conceptual Foundations of Quantum Mechanics*, Benjamin, Reading, Massachusetts.

Dynakov, M. J., and Perel, V. P. (1965): *JETP* **20**, 997.

Edmonds, A. R. (1957): *Angular Momentum in Quantum Mechanics*, Princeton University Press, Princeton, New Jersey.

Eminyan, M., McAdam, K., Slevin, J., and Kleinpoppen, H. (1974): *J. Phys.* **B7**, 1519.

Fano, U. (1953): *Phys. Rev.* **90**, 577.

Fano, U. (1957): *Rev. Mod. Phys.* **29**, 74.

Fano, U., and Komoto, M. (1977): X. ICPEAC, Paris, Abstract of Papers, p. 516.

Fano, U., and Macek, J. (1973): *Rev. Mod. Phys.* **45**, 553.

Farago, P. S. (1971): *Rep. Prog. Phys.* **34**, 1055.

Feynmann, R. P., Vernon, F. L., and Hellwarth, R. W. (1957): *J. Appl. Phys.* **28**, 49.

Gabriel, H. (1969): *Phys. Rev.* **181**, 506.

Gottfried, K. (1966): *Quantum Mechanics*, Benjamin, New York.

Haake, F. (1973), in: *Springer Tracts of Modern Physics*, Vol. 66, Springer, Berlin.

Haken, H. (1970), in: *Encyclopedia of Physics*, Vol. XXV/2c, Springer, Berlin.

Haken, H. (1978): *Synergetics*, Springer, Berlin.

Hanle, W. (1924): *Z. Phys.* **30**, 93.

Hanle, W., and Kleinpoppen, H. (1978 and 1979): *Progress in Atomic Spectroscopy*, Vols. A and B, Plenum Press, New York.

Happer, W. (1972): *Rev. Mod. Phys.* **44**, 169.

Herman, H. W., and Hertel, I. V. (1979), in: *Coherence and Correlation in Atomic Physics*, eds. Kleinpoppen, H., and Williams, J. F., Plenum Press, New York.

Hertel, I. V., and Stoll, W. (1978): *Adv. At. Mol. Phys.* **13**, 113, Academic Press, New York.

Hollywood, M. I., Crowe, A., and Williams, J. F. (1979): *J. Phys.* **B12** 819.

Jammer, M. (1974): *The Philosophy of Quantum Mechanics*, Wiley, New York.

Jauch, J. M. (1973): *Are Quanta Real?* Indiana University Press, Bloomington, Indiana.

Kessler, J. (1976): *Polarised Electrons*, Springer, Berlin.

King, G., Adams, A., and Read, F. H. (1972): *J. Phys.* **B5**, L254.

Kleinpoppen, H. (1969), in: *Physics of the One and Two Electron Atoms* (eds. Bopp, F., and Kleinpoppen, H.), North-Holland, Amsterdam.

Lamb, F. K., and Ter Haar, D. (1971): *Phys. Rep.* **2**, 253.

Landau, L. D., and Lifschitz, E. M. (1965): *Relativistic Quantum Theory*, Pergamon, Oxford.

Loisell, W. H. (1973): *Quantum Statistical Properties of Radiation*, Wiley, New York.

Macek, J., and Burns, D. (1976), in: *Beam Foil Spectroscopy* (ed. Bashkin, S.), Springer, Berlin.

Macek, J., and Jaecks, D. H. (1971): *Phys. Rev.* **A 4**, 1288.

McConkey, J. W. (1980), in: *Coherence and Correlation in Atomic Physics* (eds. Kleinpoppen, H., and Williams, J. F.), Plenum Press, New York.

McMaster, W. (1954): *Am. J. Phys.* **22**, 357.

Messiah, A. (1965): *Quantum Mechanics*, North-Holland, Amsterdam.

Omont, A. (1977): *Prog. Quantum Electron.* **5**, 69.

Pauli, W. (1928): *Sommerfeld-Festschrift*, Hirzel, Leipzig.

Percival, I. C., and Seaton, M. J. (1957): *Philos. Trans. R. Soc. A* **251**, 113.

Prigogine, I. (1981): *The Microscopic Theory of Irreversible Processes*, to be published.

Robson, B. A. (1974): *The Theory of Polarisation Phenomena*, Clarendon Press, Oxford.

Rodberg, L. S., and Thaler, R. M. (1967): *Introduction to the Quantum Theory of Scattering*, Academic Press, New York.

Roman, P. (1965): *Advanced Quantum Mechanics*, Addison-Wesley, New York.

Sargent, M., Scully, M. O., and Lamb, W. E. (1974): *Laser Physics*, Addison-Wesley, New York.

Series, G. W., and Dodd, J. N. (1978), in: *Progress in Atomic Spectroscopy* (eds. Hanle, W., and Kleinpoppen, H.), Plenum Press, New York.

Steffen, R. M., and Alder, K. (1975), in: *Electromagnetic Interactions in Nuclear Spectroscopy* (ed. Hamilton, W. D.), North-Holland, Amsterdam.

Steph, N. C., and Golden, D. E. (1980): *Phys. Rev.* **A21**, 1848.

ter Haar, D. (1961): *Rep. Prog. Phys.* **24**, 304.

Tolman, R. C. (1954): *The Principles of Statistical Mechanics*, Oxford University Press, London.

von Neumann, J. (1927): *Göttinger Nachrichten* 245.

von Neumann, J. (1955): *Mathematical Foundations of Quantum Mechanics*, Princeton University Press, Princeton, New Jersey.

Walther, H. (1976): *Laser Spectroscopy of Atoms and Molecules*, Springer, Berlin.

Zubarev, D. N. (1974): *Nonequilibrium Statistical Thermodynamics*, Plenum Press, New York.

Zwanzig, R. (1960): *J. Chem. Phys.* **33**, 1338.

Index